Robert Krueger, Hermine Vedogbeton, Mustapha Fofana and Wole S

Smart Villages

T0092534

Integrated Global STEM

Edited by
Robert Krueger and Wole Soboyejo

Volume 2

Smart Villages

Generative Innovation for Livelihood Development

Edited by
Robert Krueger, Hermine Vedogbeton, Mustapha Fofana
and Wole Soboyejo

DE GRUYTER WPI Press

Editors

Prof. Robert Krueger
Institute of Science and Technology for
Development
Worcester Polytechnic Institute
100 Institute Road
Worcester, MA 01609
USA
krueger@wpi.edu

Prof. Mustapha Fofana
Department of Mechanical and Materials
Engineering
Worcester Polytechnic Institute
100 Institute Road
Worcester, MA 01609
USA
msfofana@wpi.edu

Dr. Hermine Vedogbeton
Department of Social Science and Policy Studies
Worcester Polytechnic Institute
100 Institute Road
Worcester, MA 01609
USA
hvedogbeton@wpi.edu

Prof. Wole Soboyejo
Institute of Science and Technology for
Development
Worcester Polytechnic Institute
100 Institute Road
Worcester, MA 01609
USA
wsoboyejo@wpi.edu

ISBN 978-3-11-078621-7
e-ISBN (PDF) 978-3-11-078623-1
e-ISBN (EPUB) 978-3-11-078624-8

Library of Congress Control Number: 2023908113

Bibliographic information published by the Deutsche Nationalbibliothek
The Deutsche Nationalbibliothek lists this publication in the Deutsche Nationalbibliografie;
detailed bibliographic data are available on the internet at http://dnb.dnb.de.

© 2024 Walter de Gruyter GmbH, Berlin/Boston
Cover image: Robert Krueger
Typesetting: Integra Software Services Pvt. Ltd.
Printing and binding: CPI books GmbH, Leck

www.degruyter.com

List of contributing authors

Margaret Gunville
WPI student,
senior,
Mechanical Engineering

Christopher Martenson
WPI student,
Robotics and Mechanical Engineering

Kathryn Rodriguez
WPI student,
Biology and Biotechnology

Cameron Cronin
WPI student,
Computer Science

Connor Cumming
WPI student,
Mechanical Engineering

Aidan Horn
WPI student,
Data Science

Aaliyah Royer
WPI student,
Industrial Engineering

Lauryn Whiteside
WPI student,
Robotics

Osabarima Owusu Baafi Aboagye, III
Chief, Akyem Dwenase, Akyem Abuakwa
Traditional Area

Barima Opong Kyekyeku
Chief, Akyem Batabi, Akyem Abuakwa Traditional
Area

Barima Ahenkora
Abompe Akyem Abuakwa Traditional Area

Barima Osei Adom
Chief, Akyem Tumfa,
Akyem Abuakwa Traditional Area

Kwasi Anane Asare
Agricultural Chief,
Akyem Dwenase,
Akyem Abuakwa Traditional Area

Attah Asante
Education Chief,
Akyem Dwenase,
Akyem Abuakwa Traditional Area

Hermine Vedogbeton
Social Science and Policy Studies Department

Robert Krueger
Institute of Science and Technology for
Development

Adam Dincher
WPI student,
Architectural Engineering

Karris Krueger
WPI student,
Architectural Engineering

Casey Snow
WPI student,
Computer Science

Kelvin Atograph
E-waste processor and computer repairer,
Agbogbloshie,
Ghana

Joseph Atedoghu
ACUC student,
Engineering

Fiifi Addae
ACUC student,
Engineering

Ayomide Akerejola
ACUC student,
Engineering

https://doi.org/10.1515/9783110786231-202

Nelson Boateng
President,
Nelplast,
LtD

Grace Fitzpatrick-Schmidt
WPI student,
Chemical Engineering

Helen Le
WPI student,
Computer Science and Robotics

Ethan Wilke
WPI student,
Biomedical Engineering

Nicholas Masse
WPI student,
Computer Science

Marika Bogdanovich
WPI student,
Mechanical Engineering

Julia Leshka Jankowski
WPI student,
Mechanical Engineering and Environmental and
Sustainability Studies

Elizabeth DiRuzza
Mechanical Engineering

Sophie Kurdziel
WPI student,
Data Science

Suali
E-waste processor and computer repairer,
Agbogbloshie,
Ghana

Julian Bennett
Lecturer, Academic City University College,
Ghana

Melinda Tatenda
ACUC student,
Mechanical Engineering

Alice Abigail Tatenda Bere
ACUC student,
Mechanical Engineering

Farouk Tetty-Larbie
ACUC student,
Mechanical Engineering

Nada Abojaradeh
WPI student,
Industrial Engineering

Contents

Robert Krueger, Hermine Vedogbeton, Wole Soboyejo

Introduction

This book is about "smart villages." Having said that, it is not the kind of "smart" that you might typically read about in the literature on smart cities – or even smart villages! For example, when referring to smart cities, one's mind often drifts to information and communication technology (ICT), facial recognition and biometric technology, the internet of things (IOT), or bioinspired technology. These themes are related to, for example, e-government and e-services (ICT) in Bologna, Italy. Or surveillance, in China, for example, clothing chains, such as H&M, deploy moderate high-resolution cameras that track customers' behavior as they move through the store. These databases, which are connected to official databases, not only improve customer service but help in truly individualized marketing. The IOT can be used for on-demand city lighting, waste management, or parking. In Barcelona, street trees are outfitted with Libelium sensors that detect soil moisture and signal when a tree should receive water. In Tübingen, Germany, a parking garage uses sensors to move cars through a multistory parking garage whose only human employees are repair and maintenance technicians.

Smart city technology has entered the Global South, too. The city of Nairobi, Kenya, uses IOT technology to manage the water grid. Water customers in the city's informal settlements, who rely on distributed water supplies, before setting out, can view the "water seller grid" to determine which outlets have supply. Where there is a centralized system, the city of Nairobi can monitor the entire system and detect leaks and places where people steal water from the city's lines. In Egypt, between Cairo and Alexandria, "Smart Village Egypt" is a rural area of commerce – shops, cinemas, entertainment, and factories, that is connected via ICT and IOT. Indonesia is seeking to realize through IOT technology its "Making Indonesia 4.0" strategic plan. Smart city technology, then, is ubiquitous.

As we said at the outset, we are not talking about this kind of "smart." Is it the "village" part of the phrase that distinguishes our work? All over Europe, rural enclaves are taking advantage of "smart" technology. Like their urban counterparts, communes are using ICT to promote democracy, enhance options for local healthcare, and create car share opportunities that were once relegated to more dense areas. These rural areas are also using smart technology to sell themselves to tourists. Through smart technology, you can take a virtual tour of a village, click on a QR code to reveal detailed descriptions of cultural sites, too. These approaches have brought innovation and commerce to villages in the Global South, too. India has got in on the smart village action, too. The government and private firms are bringing high-speed internet to rural areas where college graduates return to after leaving university or their jobs in one of India's dynamic cities.

Robert Krueger, Director, Institute of Science and Technology for Development
Hermine Vedogbeton, Social Science and Policy Studies Department
Wole Soboyejo, Institute of Science and Technology for Development

https://doi.org/10.1515/9783110786231-001

This high-speed service enables IT professionals to work from where their roots, their social and familial networks, and other interests are – their home villages. Malaysia mixes these approaches into a smart agricultural system all its own. Using systems that were design in Malaysia with the support of European tech firms, farmers now have access to real-time pH and weather monitoring, pest control, and more.

At last, the technology that you will read about in the following pages is not considered "frontier" technology. In the cases that follow, you will not read about drones that deliver serum, blood, or anti-venom. You will not read about the latest refrigeration technology in support of food supply chains. And the only kind of facial recognition "software" you will hear about is the kind that comes preloaded on a human being at birth, with updates being made through the course of life experience, not periodic system upgrades or security patches.

Appropriately, the smart village technology you will read in the following pages is what we call "ordinary technology." In the cases that comprise this volume, we show how ordinary technological innovation can be "generative" because it is co-created and co-designed in communities in the West African country of Ghana. While Ghana is the site of our cases, this approach is transferable beyond Ghana and West Africa – indeed Ghana is a mishmash of borders outlined by erstwhile colonial fathers. Around the world, there are innovations that are contingent on cultural reference points to problems: gunpowder in China, the "crapper" in London, the seasonal migration of the Maasai, the weaponization of the tsetse fly by the Shona, the use of fractals to support community livelihoods, safety, and the sustainability of cultural mores.

Furthermore, it is these ordinary technologies that are going to help to fuel sub-Saharan Africa's economic independence. To be sure, hi-tech jobs are needed across the subregion. Who brought us mobile money almost a decade before Venmo and Zille? Hi-tech is *not* the silver bullet for the continent's renaissance. Without a comprehensive economic development strategy for creating off-farm jobs, sub-Saharan Africa will continue to do as it has for 500 years, and support the commodity needs of those outside its borders. It is time for the continent that is richest in natural resources and human resources to leverage their embedded competitive and comparative advantage to create sustainable societies for local communities, first, not have them part of an off-site mitigation strategy for countries and corporations in the Global North. For us, then, a smart village is one that prioritizes community-based partnerships that co-create and co-design solutions, both products and processes, to solve problems of community importance. Problems here are determined by people and their experiences, not acontextual observations from available data from certain sensors that were donated or purchased by a city administration. Solutions do not aim for some biased "mean" but account for shared understanding of community concerns and the possible solutions.

Before moving on, we need to define one additional concept: generative. Generative solutions are those that are emergent through a process that enables all kinds of voices to contribute to solutions. We also use generative to include the value that is tied to a solution and promotes a self-sustaining course that circulates through the community or region. Value is not extracted *from* the community, but generated within it, to circulate within it. Often, generative solutions are not hi-tech, but are the source of ordinary innovation that we described above. For us, ordinary solutions are also smart. We thus seek to add an additional anchor point in the "smart" discourse, in general, and the smart village discourse, in particular. We do this by valorizing ordinary technology and its innovation alongside frontier innovation and hi-technology.

This brings us to the first goal of this text, which is to release the concept of "smart" from its moorings to high tech.

The instrumented city becomes the instrumental smart city

As currently framed, the smart city leverages information and technology to provide economic opportunity, improve quality of life, and optimize city administrative functions. Smart techniques create efficiencies through large-scale data collection and analysis. Advocates for smart cities see in information technology and engineering "infinite sources of data and energy, through which cities can be managed and powered, in a sustainable manner" [1–3].

Dating back to the early 1970s, and subsidized by IBM, Los Angeles became the world's first smart city. IBM referred to Los Angeles as the instrumented city. Indeed, Vallianatos [4] shows that in 1974 Los Angeles's planning and development was being shaped by what we call today data science. In particular, the Community Analysis Bureau then employed state-of-the-art computer technologies, to process and organize huge amounts of data on different themes such as housing, traffic, crime, and poverty [1].

Today, the smart city has moved beyond basic mapping and reliance on static data. With sensors, IOT, and broadband, cities are veritable gold mines of real-time information. Large volumes of data collection, connectivity, and sophisticated algorithms allow for the creation of real time, actionable information. In this way, cities are understood not as being "data-driven," rather, they are now data-informed, meaning that cities are becoming more interlinked, which allows big data to "set the urban agenda and influence and control how city systems respond and perform" [2]. Through the integration of technologies such as sensors, grids, big-data networks, autonomous transport systems, and generators of renewable energy, the smart city ideal promises to improve the ratio of energy production/ energy waste and reduce the economic and the environmental costs generated by urban living including climate change by measuring and mitigating carbon emissions [5–8].

There are other benefits, too. Proponents of smart cities argue that improved decision-making allows for better representation, service provision, and important evidence in cases that negatively impact public safety, turning a city of complex and interwoven streets into a manageable set of tables and charts to be analyzed by administrators. E-government has given those with access to a computer to comment on projects, new ordinances, and other activities undertaken by city administrations. Public transportation and other city infrastructures will also see significant improvement from increased data collection. Knowing when and where public transport is most and least often used allows for the creation of efficient bus routes, reducing wait times and increasing overall use. Smart cities seek to improve the lives of citizens by analyzing where and when past violent crimes have occurred, and police officers are able to predict and prevent crimes before they happen [9]. For example, Tigre, Argentina, reduced auto theft and vandalism by as much as 80% with smart city technologies.

Smart cities have their critics, too. Commentators in the smart city literature have several problems related to smart cities, though too many to go into here. First, smart technology cannot overcome disparities if the monitoring and measurement techniques bias certain populations. This can have both negative and positive consequences. At their core, though, smart cities technologies, and therefore the information they provide, can, at best, only speak to the mean. If your community's problem cannot be observed through smart city techniques, or those questions are just not asked of the data, then distributive justice is rather narrowly handed out. For Kitchin [2], an international expert on smart city technique, policy, and planning, this kind of urban development enabled by technology and administrators has "ignored the metaphysical aspects of human life and the role of politics, ideology, social structures, capital and culture in shaping urban relations, governance and development." In doing so, these techniques reduce the complexity and dynamism of a city's concerns. This ignores the power relations of who defines the problems, the methods for assessing and measuring them, and the lack of broad-based engagement with society in these decisions, not to mention the technological limitations of understanding the many social, racial, and ethnic groups with their own contingent relationships to larger problems.

Do not misunderstand us, this kind of information and analysis can be very helpful. When it comes to navigating an ambulance through a morass of traffic, real-time information cannot be minimized. Smart city data can also tell that ambulance which hospital to go to with shorter wait times or where specialists practice that will support the needs of their patients. This could be especially useful to support medical technicians responding to a mass shooting in the United States.

Small is beautiful?

The smart concept has found its way out to rural areas, too. The smart village cause has been taken up by the European Union, Indonesia, Malaysia, and other states around the world. As we mentioned above, European villages have employed smart village techniques to cultivate tourism. This is one of many adaptations of the smart concept in the village context.

Like their more populous counterparts, villages around the world, from the Sahel region, to Malaysia, the United States, and Germany have adapted their needs to the smart city discourse. We offered examples of these expressions above. They fall into the same trap that Kitchin [2] mentions regarding their instrumentalization. For example, Darwin et al. [10] define a smart village as a community that is "empowered by digital technology and open innovation platforms to access global markets" [10]. Rwanda, with support from the UK, the EU, and the World Bank, has trodden along this path. In this east-central African country, the government has invested millions of pounds and euros to develop a smart refrigeration network to lengthen the marketability of locally produced foods, such as tomatoes and fish from the Lake Victoria region. Here, though, the goal has been to provide food traders with access to refrigeration so that their crops will reach markets that yield values 10 times higher than local markets. Again, this denigrates African communities as sites of extraction to support markets in the Global North. Indeed, local traders have their own innovative networks that get the food to different markets to minimize waste. For example, the high-end market is Kigali, where trucks and motorcyclists set out before daylight to bring their products to local markets. Different vendors travel to Burundi, Tanzania, and the Democratic Republic of Congo to deliver foods that appeal to those local markets. The point is that profit maximization may not be the best solution for already existing local economies.

With around 3.4 billion people globally living in villages, the smart village has become one of the greatest opportunities to expand into the emerging market of villages around the world [11]. Global corporations and brands see these communities as enormous, untapped sources of potential for economic growth. In order for these companies and industry partners to expand their markets and offer the right products and services to the villagers, they need to understand what villagers want. This ultimately allows corporations to reveal what villagers are willing to pay for. Therefore, smart village organizations are able to sell themselves as initiatives put in place to solve prevalent pain points by providing technologies along with innovative business models. This is one model.

Ideally, we believe that smart villages are intended to eradicate poverty, enhance opportunities that promote self-sufficiency and, therefore, sustainability of rural populations, and achieve development by empowering people through enduring and home-grown economic activity through digital technologies. In the chapters that follow, we layout generative, versus extractive, approaches to smart villages. We do this by relying on village expertise, listening to their problems, and deploying co-designed and co-created ordinary technologies to serve this purpose. We will now elaborate on the concept of generative smart villages.

Generative justice and the smart village and why?

In this section, we introduce the concept and practice of generative justice in smart village projects in low- and middle-income countries. Generative justice is the notion that we generate value through the transformation of materials through human labor and, further, everyone should have the right to share in the benefits of this socially produced value and nurture self-sustaining paths of circulation. The concept of generative justice shares these values with recent design theorists, too. In its best form, design meets people at where they are, co-defines problems, co-creates solutions, and imagines how an intervention's agency will shape and be shaped by the agency of living and nonliving things. Generative Justice implies new economic, environment, and social relations that move away from distributive justice to keep people close to the value they generate rather than relying exchange beyond ethnic or national borders. In doing so, we can work toward achieving a de-colonial mindset and see the vast range of opportunities that can contribute to a world that is sustainable and promotes self-sufficiency rather than continued dependence.

Using a generative justice approach in projects or designs has a great benefit for those who are designing and those who are being affected by the design. These benefits are felt when those who are affected have the power, meaning the power and decisions are made by the people living in the society. The first benefit of a generative justice approach that is talked about in this section is that it follows a bottom-up approach, which focuses on the community coming up with ideas or even the problems rather than being a "higher up." Next is the effect that this approach has on the ecological sector or the environment. This is a positive benefit at least in the view of the community it affects. Another benefit occurs in the labor/economic sector where there is a different focus on who does the labor and who economically benefits from the labor. The last benefit of generative justice discussed is with the social and political sector which comes from people being in charge of decisions for themselves and their community.

A bottom-up approach offers sharp focus on the people who are closest to the site of decision-making and intervention. Bottom-up approach provides a contrast to top-down processes. Top-down decision-making gives the power to decide to elites, experts, financiers, and donors. The bottom-up approach supports the creation of bespoke interventions, for a community, rather than relying completely on technology transfer or mobility. If exercised with a generative justice sensibility "value generated by labor and/or nature are to be governed by the people and recirculated within the systems doing the producing, rather than redistributed by a centralized state" [12]. The circulation of value is at the root of where the people are, which gives communities a say in what is going on in their society. For Eglash [13], this helps allow freer expression, individuality, "address wealth inequality and environmental degradation." This approach, as you will see, empowers groups who have been disenfranchised for centuries. As one of our partners said, "I never knew how to build a bridge. Together we built one, now I can do it on my own." This is the way we cultivate self-sufficiency and, therefore, the sustainability of our projects and interventions.

Generative justice also valorizes different, or nontraditional, sources of value. Generative justice supports this broader sense of value. Value here extends beyond the exchange value or market transactions. For Dotson and Wilcox [12], the value is typically focused on narrow factors such as "electrons and dollars [which] distracts from the broader range of values currently not being generated and recirculated within a community." On a recent trip to Akyem Abuakwa, in the eastern region of Ghana, His Majesty, The Okeyenhene, Osagyefuo Amoatia Ofori Panin, remarked to our team of faculty and student visitors that, more than economic value, people in the region he represents need to see "things" that they identify these with progress. Seeing this creates a sense of hope among the community. When they are involved in the design and development, this has a multiplier effect. It can be a bridge, a funeral ground, a toilet, or a new bore hole. Offering this sense of hope can lead to new values that reflect soil and water quality, ecosystem health, and community self-sufficiency.

Generative justice includes communities as partners, peers, and experts. They are not diminished because of how they have learned or their livelihoods. Through the process of collaborative problem-solving, co-design, and co-creation, they learn their own agency in bringing about community change. Ghana's first president, Kwame Nkrumah, suggested in a speech on his notion of Pan Africanism that "socialism is the economic system that most resembles African community economies." He was both right and wrong. His socialist policies, like those of Mugabe, Machel, and Nujoma, demonstrated that Soviet-style socialism was not fit for African any more than it was for the USSR. Where he was right is that African communities valorize community cohesion, shared interests, and solidarity. Perhaps this is where the concept of smart villages in Africa is most resonant, and where other parts of the world should take notice as we adapt to climate change and other social disasters.

Some final thoughts

Western and non-Western communities often showcase different cultures and values. These often stark differences are manifested in developmental goals, such as modernization, that remain pervasive in the relationship between low- and middle-income countries and industrialized countries. Western communities are often seen as individualistic, valuing large infrastructure and a consumerist-free market with a time orientation to the future [14]. On the other hand, non-Western communities are often collectivistic, valuing a community-oriented structure that depends less on material items [14]. These societies have sought different development trajectories based on their values. This stance has resulted in them being seen as traditional, backward, and inferior. The arrogance of the west and the forced capitulation of former colonies have created severe cultural misunderstanding, arbitrary policy and practice, and massive destruction of the way of life of almost a billion people. The time has come for development goals and interventions to relate to unique cultural values of a place, and to eschew the notion that one-size-fits-all development is the path to global modernity.

The chapters

The subsequent chapters in this book were originally written, with support from Vedogbeton and Krueger, by WPI students working on their junior project which is ideally designed to help students understand the relationship between science, technology, innovation, and society. Some chapters were written with WPI seniors working on their senior theses.

The chapters are intended to be read in a number of ways. First, if you are an educator, you could read the following chapters as methods for developing more equitable and inclusive approaches to teaching STEM (science, technology, engineering, and mathematics) courses, especially those with a design component. As an educator, you may also draw inspiration from the chapters as you link the ideas contained here to your own scholarship, innovation, teaching, and practice. Students, as well as educators, could also read these chapters with an eye toward self-examination, especially around implicit bias. Implicit bias affects how we see "others." Others can be people like BIPOC, nonbinary, or women in STEM. Others can also be forms of inquiry, such as those influenced outside Western culture. Africa, for example, has a long history of science, technology, and innovation, but has been excused from the canon as not up to the standard because it is recorded and passed along verbally through generations. It was coopted by colonial scientists and military strategists, or just not seen because our way of framing knowledge is so vastly different that we miss it. Finally, others can be some combination of both of these. For example, what qualifies as expertise and who has it? This list is not exhaustive, and we invite you to come up with your own ways of reading and learning from these collaborations. Also note that there is a long list of authors on each chapter. We recognize the importance of our Ghanaian partners and include "their" knowledge in the texts. All too often Western researchers and practitioners ignore the integration of knowledge that we produce through the process of problem definition, the co-design of solutions, and the co-creation of the final intervention. Worse yet, researchers and practitioners work with the assumption that they know best and do not need input from local partners. If you are one of these people, we hope that the chapters that follow will help you see things in new ways.

Chapter 1 begins our presentation of the themes discussed above by advancing the book's thesis on generative justice by demonstrating that co-designing projects with the first users' needs in mind can sustainably help address issues that affect communities around the world. In "Burn Box," the authors collaborate with workers at the Agbogbloshie e-waste site, the largest in Africa, to reduce air pollution that has a negative impact not only on e-waste wire burner but also on the community at Agbogbloshie and its surroundings. Wire burning removes PVC coatings to reveal copper, which is the source of income for many workers at Agbogbloshie. Indeed, the government of Ghana has sought to stop this practice because of its deleterious effects on human health and the environment. Finding a solution to reduce air pollution is critical to sustain workers' livelihood. The authors partnered with students from Academic City University College (ACUC) in Ghana to design an incinerator (i.e., burn box) and a scrubber to reduce pollution from wire burning. The design is equipped with a filtration system that retains pollutants that would have otherwise been released to the environment. The co-design process relied on an understanding of the main steps involved in the extraction of copper from electronic wires, and the health effect of exposure to air pollutants. The partners also completed ethnographic work to understand the routes of exposure that are prevalent at the site. This includes women, children, and the male waste workers. The partners used e-waste materials that are found at the e-waste site to create a burn box with a technology that is easily replicable and affordable. Understanding this community issue and working with them to generate ideas that helped develop a burn box demonstrate the type of generative justice we described above.

Chapter 2 presents another case of generative justice where knowledge is co-created to empower our partners in Ghana to create their own technology out of waste. In "Stirling engine," the partners sought to address an unmet community need: access to electricity for charging cell phones. Remember that the average wage in Accra is around $300/month, with electricity costs per kWh reflecting a much more significant cost based on per capita income than in the United States. The partners reverse engineered a Stirling engine to produce electricity using e-waste materials available at Agbogbloshie e-waste site. The engineering design process relied on selecting and replacing the Stirling engine's components with e-waste materials. The resulting product is a Stirling engine that can be used to charge cell phones in areas with no access to electricity. It can be reproduced by Ghanaian partners who have been involved in all the design and creation process.

In Chapter 3, the partners sought a longer term view to generative justice: to develop a recycling economy based on Akan tradition and values. In "Generative Recycling System," the partners developed a community-based plastic waste recycling program to help villagers properly discard their waste, and, more importantly, see plastic waste (high-density polyethylene, low-density polyethylene, and polyethylene terephthalate) as a raw material for new products, such as building materials. This first exploratory case study involved the villages of Dwenase, Tumfa, Abompe, and Batatbi, which are all located in the Kingdom of Akyem Abuakwa. Led by Chief Osabarima Owusu Baafi Aboagye, III, the partners included traditional leadership from the four towns. In consultation with the four communities, the students proposed a recycling economy that would capitalize on the material value of plastic waste and is self-sustaining and mutually profitable to all the villages involved in the project.

To demonstrate the value of used plastic, which is ubiquitous in the region, the partners designed an irrigation demonstration site using blocks made from recycled plastic. In "utilizing recycled plastics," the partners designed a drip irrigation system to supply water to 16 acres of cocoa trees to support Akyem Dwenase's economy. Climate change has affected the rain patterns across equatorial Africa. Over the past few years, rains either come early or late. The irrigation system will help crops that have not yet matured to live during periods of rainfall variability.

Finally, in Chapter 4, the authors collaborated with nonworking women in Agbogbloshie to bring awareness on the effect of e-waste pollution on their health. The initial project was to find solutions to eliminate e-waste pollutants found in water, food, and soil. However, through their interaction with the women, the students changed their focus from finding solutions to mitigate pollution to bringing awareness on the negative effect of pollutants on their health and those of their children. Understanding women's daily routine and the types of contaminants on the e-waste site support the student's presentation of the health-related problems to these pollutants.

References

[1] Cugurullo, F. (2018). The origin of the Smart City imaginary: from the dawn of modernity to the eclipse of reason. In Lindner, C. & Meissner, M., editors. *The Routledge Companion to Urban Imaginaries*. London: Routledge.

[2] Kitchin, R. (2014). The real-time city? Big data and smart urbanism. *GeoJournal*, *79*(1), 1–14.

[3] Taylor Buck, N. & Aidan, W. (2017). Competitive urbanism and the limits to smart city innovation: The UK future cities initiative. *Urban Studies*, *54*(2), 501–519.

[4] Vallianatos, M. (2015). *Uncovering the Early History of "Big Data" and the "Smart City" in Los Angeles.* Boom California. https://boomcalifornia.com/2015/06/16/uncovering-the-early-history-of-big-data-and-the-smart-city-in-la/.

[5] Garau, C., Masala, F., & Pinna, F. (2016b). Cagliari and smart urban mobility: Analysis and comparison. *Cities*, *56*, 35–46.

[6] McLean, A., Bulkeley, H., & Crang, M. (2016). Negotiating the urban smart grid: Socio-technical experimentation in the city of Austin. *Urban Studies*, *53*(15), 3246–3263.

[7] Vanolo, A. (2014). Smart mentality: The smart city as disciplinary strategy. *Urban Studies*, *51*(5), 883–898.

[8] Viitanen, J. & Kingston, R. (2014). Smart cities and green growth: Outsourcing democratic and environmental resilience to the global technology sector. *Environment and Planning A, 46*(4), 803–819.

[9] Joh, E. (2019). Policing the smart city. *International Journal of Law in Context*, *15*(2), 177–182.

[10] Darwin, S., Fischer, W., & Cheesborough, H. (2020). *How to Create Smart Villages: Open Innovation Solutions for Emerging Markets*. Peaceful Evolution Publishing.

[11] Darwin, S. N., Chesbrough, H., & Rotter, K. B. (2018). Prototyping a Scalable Smart Village (B). In *SAGE Business Cases*. The Berkeley-Haas Case Series. Berkeley: University of California. Haas School of Business.

[12] Dotson, T. C. & Wilcox, J. E. (2016). Generating community, generating justice? The production and circulation of value in community energy initiatives. *Teknokultura*, *13*(2), 511–540.

[13] Eglash, R. (2016). Of Marx and makers: An historical perspective on generative justice. *Teknokultura*, *13*(1), 245–269.

[14] Mavhunga, C. (2017). *What do Science, Technology, and Innovation Mean from Africa?* The MIT Press.

Margaret Gunville, Christopher Martenson, Kathryn Rodriguez

Chapter 1
Burn box case study

1.1 Introduction

Many higher income nations such as the United States and the states within the European Union (EU) have made it illegal to dispose of electronics in landfills but continue to export e-waste to lower income nations that do not have the resources to reject the imports [1]. As a result of the poorly enforced e-waste regulations in these developed nations, the majority of 59 million tons of e-waste generated in 2019 was illegally exported to low-income countries [2].

Electronic waste or e-waste is composed of discarded obsolete electronic devices. At e-waste sites, recycling practices have become a type of informal economy. Components from electronic devices are either fixed or recycled for their raw materials. This includes selling the inner copper core from wires brought in from all over the sites. E-waste recycling is an important facet of the economy but has a serious health hazard impact on recyclers [3].

Wires are burned at e-waste sites to remove the plastic coating, which presents a significant health hazard. The burns release a thick gray ashy smoke that contains toxic heavy metals, hydrochloric acid (HCl), and halogenated organic compounds, which pollute and bioaccumulate in the waters and soils of surrounding areas. Direct exposure to these emissions occurs through inhalation and ingestion, posing a significant threat to human health [4]. Many of these severe health effects are still understudied; however, chromosomal abnormalities, neurotoxicity, and carcinogenic effects have been observed [5, 6]. This study focuses on Agbogbloshie, an e-waste site in Accra, Ghana, where unregulated recycling activities pose a major health threat to recyclers and their communities.

In this case study, we collaborated with the community in Accra to design a burn box to allow for the safe burning of wires with limited emissions to reduce e-waste pollution health hazards. We considered the specific user needs, materials, and means available to repair and reproduce our design. By co-designing and taking the aforementioned variables into account, we aimed not only to meet the needs of the Accra community but also to build a foundation to allow for self-sufficiency and sustainability in the long term.

Technology without the input of the end users can lead to unintended consequences. If left out of the design process, communities may be solely dependent on the de-

Margaret Gunville, WPI student, senior, Mechanical Engineering
Christopher Martenson, WPI student, Robotics and Mechanical Engineering
Kathryn Rodriguez, WPI student, Biology and Biotechnology

https://doi.org/10.1515/9783110786231-002

signer. A technology reliant on the designer is inherently short term as it prevents communities from creating a sustainable foundation to repair and build upon a technological solution. Our approach is different from the past literature because we collaborated with our partners in Accra and Agbogbloshie, Ghana, to co-design a functioning and tested prototype burn box equipped with a filtration system. There are many hazards associated with wire burning; hence, the process is currently illegal. We hope that our combined effort in designing the burn box will allow for the safe burning of wires with limited emissions, which will help to legalize burning and reduce strain on the community.

In the following sections, we discuss in greater detail e-waste both in the Global North and lower income countries, the history of Agbogbloshie, and issues related to the recycling of wires there. We also discuss different prior incinerator and scrubber designs to build a foundation for our design. Then, we discuss the process of co-design and how we plan to use it throughout our project. Next, through interviews, design matrices, and experiments, we will begin to design the burn box that meets the needs of wire burners while reducing the amount of pollution expelled. We will then discuss the varied materials and design options with the community. This will include the physical problem, weather patterns, and social aspects that may impact our design. By designing it with the community, we can make this an experience of developing self-sufficiency and sustainability. To ensure the viability of our prototype, we will conclude our research by comparing the emissions released from a burn within our prototype to the emissions released in a simulated open-air burn.

1.2 Literature review

E-waste has become one of the fastest growing waste streams in the world [7]. With the continuous advancement of technology, there is always a new product that is being heavily marketed to consumers who are more frequently purchasing, upgrading, and discarding their electronic products [8]. According to the U.S. Bureau of Economic Analysis, "Americans spent $71 billion on telephone and communication equipment in 2017, nearly five times what they spent in 2010 even when adjusted for inflation" [9]. The constant demand to reach a more "modern" world has created a boom in communication technologies like computers, phones, televisions, and other household items [10]. The issue with the rapidly growing procurement of innovative technologies is that when something new is purchased we get rid of what is old. Experts are claiming that this will result in a dramatic increase in e-waste [9]. Another factor that contributes to frequent upgrading is the decreasing life span of electronics. For example, the average life span of laptops has decreased from 4 to 2 years, a 50% decrease [11]. Thus, the amount of e-waste and associated health risks produced every year will also continue to increase.

No federal regulations currently require e-waste recycling in the United States. This has forced states to establish their own laws and regulations regarding e-waste. Some examples of state regulations are "(laws that require) manufacturers of computers, computer monitors, laptop and portable computers, and televisions to provide recycling services throughout the state at no cost" to the consumer. More specifically, "the State of California has passed a law charging consumer fees, called advanced recycling fees (ARFs), at the time products are purchased" [8]. While these laws do help make it easier for consumers to recycle their e-waste, most of the waste still ends up in developing countries. Eighty percent of the e-waste generated in the United States becomes part of the "hidden flow," which means that it is unofficially exported or dumped into landfills [11].

In Europe, < 40% of e-waste produced is recycled and more than half of the waste produced is large household appliances [12]. In the EU, recycling practices vary by country. Although illegal under the Basel Convention, rich countries, like many in the EU, still export unknown amounts of e-waste to lower income countries [13]. One of the main ways the EU is working to reduce its environmental footprint due to e-waste is by passing legislation that prevents certain chemicals, like lead, to be used in the makeup of electronics. Additionally, in March 2020, the European Commission adopted a new circular economy action plan. Since the initial adoption, more initiatives have been added to the original plan. The major aims of the action plan are to "make sustainable products the norm in the EU" while focusing primarily on sectors with the most waste-like electronics [14]. While plans set in place aim to reduce the amount of waste produced by the EU, exportations of the waste are still occurring.

Exporting e-waste to developing countries is common for the Global North. Labor costs and environmental regulations for hazardous waste disposal make exporting the waste the most efficient method. E-waste can provide economic benefits to lower income countries, but they are often short-lived, as these countries often lack the "technology, facilities, and resources needed to properly recycle and dispose of e-waste." This then creates public health and environmental concerns [11].

1.2.1 E-waste in low-income countries

Due to unregulated exports, it is unclear how much e-waste is exported every year. According to Perkins et al. [11], of the 20–50 million tons of e-waste generated annually, it can be estimated that 75–80% is shipped to Africa and Asia to be "recycled." This exportation of waste ends up creating villages, economies, and communities. These villages are littered with waste and exposed to health and environmental risks, but they have found a way to make their lives there [1].

E-waste recycling is a form of survival in informal urban areas of lower income countries. These sites consist of homes, restaurants, vendors, street trading, and even have their own forms of government [10]. Each site is different in the way that its economic and social structures are established, but some aspects, like health concerns, are the same across locations. Recyclers in sub-Saharan Africa work to repurpose functional electronic components, and strip anything else to remove the circuit boards and wires for the metals they contain.

A key commodity is copper. Recyclers have found that the most direct and quickest way to get the copper from inside wires is to burn the PVC coating off. While this might be an effective method, it releases highly toxic chemicals that threaten the environment and human health. During the rainy seasons, toxins from e-waste sites run off into agricultural lands, contaminating crops and farm animals that are later consumed by humans [1]. Around e-waste sites, pollutants "are found at significant doses in human serum, blood, hair, placenta, breast milk, and umbilical cord blood, indicating that exposure to e-waste presents a risk for the present as well as for future generations" [10]. Health concerns are extremely important, but for those living on or near the site, there is a need to find a way to survive, and becoming a recycler is often the answer.

1.2.2 Agbogbloshie

Agbogbloshie, located in the outskirts of Accra, Ghana, is one of the largest e-waste sites in the world. The Odaw River borders most of the e-waste site, which is approximately 0.4 km^2 in area. An image of wire burning alongside the river at Agbogbloshie is shown in Figure 1.1. Agbogbloshie is home to 80,000 people, many of whom migrated from other parts of Ghana and surrounding countries. One reason for the influx of people was the Konkomba–Nanumba conflict in Northern Ghana in 1994 [15], which saw many Ghanaians move to Old Fadama, an informal settlement near Agbogbloshie [3]. Numerous low-income families have continued to relocate to the Old Fadama and Agbogbloshie area due to the affordable housing and opportunities for unregulated or informal jobs . With shipments of e-waste from outside countries increasing over the past 10–15 years, e-waste recycling in Agbogbloshie has expanded immensely into an informal economy. This economy provides benefits not only to those in Agbogbloshie but also to all of Ghana through their e-waste efforts.

1.2.2.1 Agbogbloshie's informal economy

To understand the role Agbogbloshie plays in Ghana's e-waste economy, it is important to explain what an informal economy is and how they arise. The International Labor Organization (ILO) defines an informal economy as a "non-structured sector that has emerged in urban centers as a result of formal sectors' inability to absorb new entrants" (ILO) (1972, 9). From this definition, it is evident that the creation of an informal e-waste economy in Ghana is based on the ineffectiveness of the formal sec-

Figure 1.1: Agbogbloshie's wire burning site.

tor. The formal waste sector in Ghana deposits only 10% of solid waste into landfills and lacks the infrastructure for glass and plastic recycling [53]. ILO later defined informal economies as "consisting of units engaged in the production of goods and/or services with the primary object of generating employment and incomes to the persons concerned" [54]. Agbogbloshie's informal e-waste economy also follows this definition by providing opportunities for many workers who otherwise have little to no alternatives for jobs. The Global North refers to the Agbogbloshie's e-waste industry as informal without having a full understanding of the activities and interactions that take place. While Agbogbloshie and other informal sectors may seem foreign to the Global North, Agbogbloshie's e-waste system is far more complex than what some may believe, and it can be the best way to resolve the unemployment and waste problems in Ghana.

The informal e-waste economy in Agbogbloshie can be split into diverse types of professions that interact and trade with each other to create a profitable and productive system. In a film about Agbogbloshie, a scrap dealer named Mohammed Seidu referred to this system saying, "We are the burners, but we have others who scout in bushes to find copper, they bring it to sell to the business people, so they can bring it to our company for them to be dismantled and gather the burning copper. They bring it to us to burn." One of the job groups collects the scrap e-waste from all around the city and brings it back to Agbogbloshie to sell to intermediaries or scrap dealers. These collectors tend to be young as this is commonly an entry-level position for many who look to join the e-waste economy. Another section of the market consists of recycling printed computer boards (PCBs) and copper wires for the valuable metals (copper, gold, and aluminum) that can be extracted. Many of these resources require expensive or time-consuming methods to obtain the materials safely and effectively from e-waste. In Agbogbloshie, these methods are unavailable causing many recyclers to use less safe, but effective, methods such as burning wires in open fires to remove the plastic coatings from the copper. Refurbishers also make up the informal e-waste economy by repairing used electronics through reused e-waste parts.

With 28.5% of the Ghana population living below the poverty line [55], this market allows affordable access to computers and other electronics to a substantial number of Ghanaians. The refurbishing sector has continued to grow within all parts of Ghana with many shops employing apprentices to teach future generations about their trade. The Ghanaian government contracted shops in the informal refurbishing sector in 2009 to supply the country with laptops under the "one laptop per child initiative". In 2015, 25% of computers in beneficiary schools and households were repaired and refurbished in Agbogbloshie [52].

Another bridge from the informal e-waste market to the formal market is through scrap dealers. Scrap dealers work for urban companies out of the nearby port city Tema and are the highest position in the Agbogbloshie network [56]. The Agbogbloshie e-waste sector also includes shops that operate as intermediaries among the collectors, recyclers, and scrap dealers. Agbogbloshie also houses a substantial number of non-e-waste shops that provide equipment and consumables to the e-waste workers. These trade shops tend to be female dominated and create another market for job opportunities from the e-waste economy.

Agbogbloshie's informal e-waste market not only stimulates the Ghanian economy with job opportunities but also provides them with wages far higher than the country's minimum wage. With inconsistent financial return from agriculture and a saturation of workers in other low-skill fields, the e-waste economy can provide more financial stability for many workers and their families. Amankwaa [56] surveyed e-waste workers from Agbogbloshie and found that even the collectors, the lowest earning group, had a daily income more than three times greater than the national daily minimum wage (see

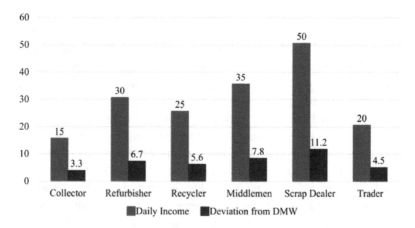

Figure 1.2: Daily income distribution of e-waste workers in Ghana (Amankwaa, 2012).

Figure 1.2, [56]). The e-waste market accounts for the income of up to 201,600 people [57] and contributes US $105–$268 million to the Ghanian economy [58].

The Agbogbloshie's e-waste economy is therefore essential to Ghana, and, if some environmental and health issues are resolved, it can become a sustainable industry for much of Ghana's workforce.

1.2.3 Wire burning

Open waste burning is prevalent throughout countries that lack the infrastructure for the proper disposal of waste. Often, to reduce the volume of waste, residential open burns and landfill burns will occur [16]. These burns are unregulated and occur at low temperatures, operating at 500 °C or lower causing unnecessary waste products to be expelled, while commercial incinerators work at temperatures of greater than 850 °C [17]. In the case of landfills, sometimes these burns are unintentional and therefore uncontrolled. All types of open burns, however, pose a significant risk to communities and the planet. An example of wire burning is shown in Figure 1.3. It was predicted by the World Bank that by 2050 waste products would reach 3.4 billion tons a year [18]. Since 41% of that waste is burned annually, 1.3 billion tons of that waste will be burned, producing soil, water, and food contamination and greenhouse gas emissions [4, 16].

Figure 1.3: Wire burning in Agbogbloshie.

Wire burning presents both economic opportunity and severe health hazards. As much of the e-waste products of the Global North are exported to low-income countries, a market for the recycling and selling of these products has been created. Wires are scavenged from old electronics, burned to remove the outside coatings, and the copper is sold in this new informal economy. In Ghana, this network of "informal employment" represents 66.7% of all employment [3]. While this does constitute livelihoods for many people, it is unsafe for the workers and their communities. The burning process unintentionally leaches harmful chemicals into the environment.

The different composition of wires impacts their burning and by-products. Wires are either insulated or uninsulated. Insulated wires are typically composed of a copper inner core surrounded by a semiconductor and covered in a PVC sheet. Burning insulated wires produces approximately 100 times the organohalogen toxins as burning municipal solid waste (MSW) [16]. While PVC is inherently flame retardant due to the chlorine atoms in its structure, some wires are also coated in brominated flame retardants (BFRs) [19, 59]. Cables are composed of multiple types or sizes of wires, usually between 10 and 14 gauge, wrapped together including grounding and inert. These inner wires are also coated in a BFR or paper coated in oil for insulation. These outer layers are broken down into combustion products when the wires are burned leaving behind the inner copper wire, ash, and smoke, which contains organohalogens, HCl, and toxic heavy metals.

1.2.3.1 Toxins and hazards of wire burning

Halogenated organic compounds constitute one of the main sources of toxic emissions from wire burns. Halogenated compounds contain elements from Group 7 of the periodic table (i.e., fluorine, chlorine, bromine, iodine, etc.), chlorine being a vital component of PVC [19]. When PVC breaks down, especially at the lower temperatures constituted by the open burns, by-products are created due to incomplete combustion [17, 20]. It has also been found that copper acts as a catalyst when thermogenically interacting with the PVC to form halogenated compounds as well as copper complexes. These by-products are found in ash residue and soil samples from surrounding areas (Fujimori et al., 2016). These halogenated compounds are known to contain dangerously high amounts of chlorine and bromine from the PVC and BFRs. Some of the most common types are polychlorinated dibenzodioxins/furans (PCDDs/Fs), polybrominated dibenzodioxins/furans (PBDDs/Fs), and polychlorinated biphenyls (PCBs) [16]. Since chlorine and bromine have such similar structures, conversion will occur between the species through substitution in the burning process [21]. These compounds can be grouped together as dioxin-related compounds (DRCs) and can be expected whenever both chlorine and bromine donors exist (Fujimori et al., 2016; 21).

DRCs constitute an ongoing threat to global health. PCDDs/Fs constitute 2 of the 12 initial persistent organic pollutants as listed at the Stockholm Convention. This was due to their ability to travel long distances, possess high toxicity, and the ability to bioaccumulate [21]. While not initially listed, other DRCs have similar physicochemical properties that make them comparable or even more toxic than the PCDDs/Fs [22]. DRCs are ubiquitous pollutants detected in the soil, air, sediments, birds, marine species, fish, house dust, and human tissues surrounding waste sites [23]. In these areas, PCDDs/Fs have been discovered in adult human tissues at 25 and 11 times the tolerable daily intake [24]. This is detrimental as DRCs have negative effects on reproductive viability and can be passed through both the placental barrier and through breast milk to developing fetuses and infants [6, 23, 25]. DRCs have been shown to cause acute neurotoxicity, especially within young children, aged 2.6–3 [25]. These neurotoxic effects show detriment to development of overall cognition, intelligence quotient, memory, language, gross and fine motor skills, attention, executive function, and behavior [6]. While there is an apparent lack of research into the chronic effects of these chemicals on human exposure, it has been suggested that these effects may be due to a combined result of thyroid hormone disruption and direct interference with neuronal development [23]. Many DRCs are also carcinogenic agents, which can damage DNA causing chromosomal aberrations and genetic mutations, increasing the lifetime risk of cancer by approximately 40% [5]. Their increased presence in the bloodstream is correlated with increased cases of non-Hodgkin's lymphoma, Hodgkin's disease, leukemia, and lung cancer [16]. Aside from releasing these organohalogen pollutants, the burning of e-waste also threatens human populations by releasing HCl.

The complete combustion of PVC releases HCl as corrosive and toxic vapor that dissolves in water and produces both acute and chronic effects [16]. There are two main classifications of toxic products released in the burning of different materials, irritants, and narcotics. HCl can be classified as an irritant [20]. The acute effects include respiratory distress caused by corrosive effects on mucous membranes and tissues. This can result in scarring and ulceration of the respiratory and digestive tracts. These chronic effects are an extension of the acute effects causing continuous degradation of the mucosal membranes [26]. Eventual effects include development of glaucoma and cataracts [16]. So far, no carcinogenic effects have been observed due to the inhalation or ingestion of HCl. Although the toxic effects of HCl do not cause generational damage, their effects are still severe and should be prevented with the other toxins released in wire burning.

Toxic heavy metals, like HCl, cause both acute and chronic effects. The main heavy metals given off in wire burning are copper (Cu), zinc (Zn), and lead (Pb) (Fujimori et al., 2016). Cu and Zn are both important to support the brain function and reaction catalysis; however, when they are found in high concentrations, they can cause acute neurotoxicity. Excess copper in the brain has subsequently been linked to genetic diseases such as Alzheimer's and Wilson's diseases; however, these have not yet been studied in conjunction with the ingestion of heavy metals. Zn is increasingly studied since it is one of the main therapeutic targets for understanding Alzheimer's, as its apparent effects include disrupting metabolism within the brain and increased Zn flux through transporters and synapses. These metals can convey their toxic effects from pregnant mothers to their fetus and therefore cause generational damage [27]. Fly ash released in the burning of e-waste, such as wires, has been found to have 200 times the acceptable Pb limits within the United States (Gullet et al., 2007). Pb, unlike Cu and Zn, is not necessary for the body to function. Pb, in amounts as small as 10–20 µg/dL, especially in children, can have devastating effects across several organ systems such as the immune cells, kidneys, and nervous system. Together with the effects of other metals, these can cause lifelong neural and physical damage [28]. Toxic ash creates an unseen threat to wire burners and the surrounding community, but there are other safety concerns that also add to the dangers of this recycling process.

The process that e-waste goes through to be recycled from wires contained within expired products to copper being sold in markets is arduous and innately physical. Electrical components will fall on people's heads. Cuts, scrapes, and burns are not accidents but part of the trade that is sustained daily [3]. Wire burners are specifically subjected to temperatures of up to approximately 500 °C for hours at a time, leading to dehydration and the potential for third-degree burns [17]. The ash from this fire, aside from its toxicity, will get into the workers' eyes and lungs, drying them out. A summary of all the physical and toxic effects of wire burning is shown in Figure 1.4. We hope to limit both these challenges to wire burning for wire recyclers and their communities by designing our burn box.

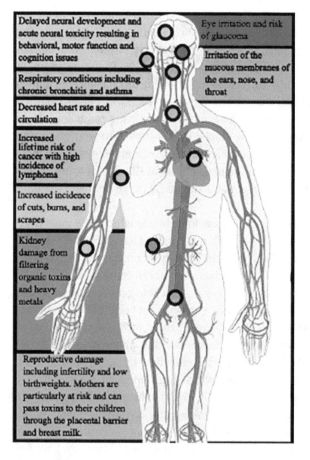

Figure 1.4: Effect of wire burning on the human body.

1.2.3.2 Wire burn fuels

The fuels used to ignite and sustain the wire burns also contribute to the toxicity of the wire burn process. Differing fuels cause variance in the burning temperatures obtained that can cause more toxic emissions from the wire itself [17, 20]. Unconventional to our perspective, matches are used to light refrigerator insulation/foam and tires as fuel for these fires [3]. These fuels, however, do not allow for even distribution of heat and give off their own toxic emissions. Fuel sources frequently used in the Global North are kerosene, propane, and liquified petroleum gas (LPG), which are less toxic and more easily regulated; however, as we explore later, these may not be appropriate for our partners in Ghana [29].

Refrigerator insulation is often composed of polyurethane, a highly flammable compound that releases toxic gases when burned. The emission profile from polyurethane depends on the temperature at which it is burned. Lower temperatures produce more smoke and heavy organic compounds, while higher temperatures create cyanide (HCN) and carbon monoxide (CO) in much larger quantities [30]. Because of the flammability of polyurethane, BFRs are also occasionally present in refrigerator insulation; however, these were already considered with the toxicity of wire burning (Fujimori et al., 2016; 19). Another major concern of using any refrigeration equipment is chlorofluorocarbons (CFCs). A molecule that reacts with ozone in the atmosphere causes damage to the ozone layer [31]. This not only causes issues for our partners in Ghana but also poses ongoing problems for the entire world.

Tires are already used as a fuel source in multiple processes in Ghana as a cheap, easily accessible alternative to firewood. It is even used directly to singe butchered meat. Tire burns often occur using tires cut into pieces or shredded. In both cases, a significant amount of organic toxins with mutagenic/carcinogenic properties are released, three to four times the amount seen from other fuel sources [32]. In addition to these organic toxins, both irritants and narcotics are released, causing irritation to the mucous membranes and respiratory issues. Of the narcotics, CO and benzene are the most notable and seen in higher concentrations when compared to the emissions of both firewood and LPG [29].

Firewood and coal are two biomass fuel sources currently used in Ghana. These fuels are readily available; however, they are most often used as fuel sources for cooking rather than wire burns. While these fuels produce less carcinogenic effects than tire and refrigerator insulation, they still produce significant amounts of irritants and narcotics such as CO and benzene, which serve to pollute the atmosphere [29].

Using a more widely accepted and easily moderated fuel source may be beneficial in our design. Gas fuels can be regulated using nozzles, which allow for a steady application of heat to be added to the furnace, while not adding other potential toxins other than those normally found in combustion reactions like CO [29]. These fuel sources, though, are more expensive and there would be no way of preventing the usual fuel sources from being added to the burn box. As listed in Table 1.1, these accelerants must be considered when designing both the box and the filtration system.

Table 1.1: Wire burning fuel summary.

Material	Burning temperature (°C)	Main toxins
Wood	600	CO, benzene, nitrogen oxides, volatile organic compounds, formaldehyde, and particulate matter
Charcoal	1,100	CO, benzene, heavy metals, sulfur dioxide, and volatile organic compounds
LPG	1,900	CO and benzene
Refrigerator insulation (polyurethane)	800	HCN, CO, CFCs, and heavy organic compounds
Tires	1,100	CO, benzene, and carcinogenic organic toxins

1.2.4 Prior art

In this section, we analyze assorted designs for small-scale incinerators and scrubbers. Components of each design serve as a reference for what may be included or needed in the final co-designed product.

1.2.4.1 De Montfort University low-cost medical waste incinerator

This incinerator design created by De Montfort University was aimed to be used in developing countries to dispose of medical waste. The incinerator is constructed to operate at 800 °C and to handle 15 kg/h of waste. While this design was not created for e-waste, the similar parameters and affordability of the design and its materials can be a reference for other waste incinerators.

The general incinerator design consists of a loading door, chimney, air inlet, ash door, and a grate (a diagram is shown in Figure 1.5 and a picture of the completed design is in Figure 1.6). The materials used are primarily fire bricks, standard building bricks, steel tubing, and sheets. This design can use both kerosene and wood as fuel for the fire.

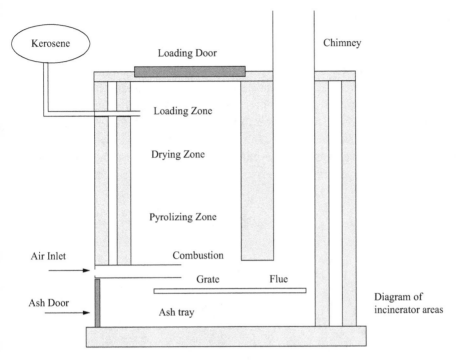

Figure 1.5: Incinerator diagram.

Figure 1.6: Incinerator completed.

1.2.4.2 Fluidized bed waste incinerator design

In Europe and North America, fluidized bed waste incinerators are used to dispose of waste. The incinerator system can separate metal for recycling and reuse the leftover ash for building materials. The steam created is utilized to create electricity by spinning a turbine. Figures 1.7 and 1.8 show different diagrams of the fluidized bed waste incinerator to help visualize the process. In Figure 1.7, it shows the inputs and outputs of a fluidized bed waste incinerator. This modern design of an incinerator shows not only how far development in the technology has come but also the major features that even complex incinerators have continued to utilize over years of development.

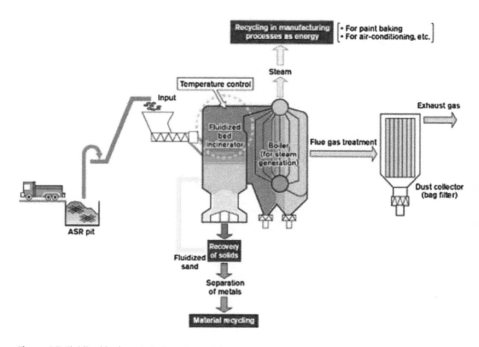

Figure 1.7: Fluidized bed waste incinerator system.

1.2.4.3 Catalytic converters and reburn tubes

Emissions from incinerators and burning stoves can be reduced by using an extended burn process. Both catalytic converters and reburn tubes are used for the burning of smoke and particulate matter given off in various combustion processes. There are two main types of catalytic converters: two-way and three-way. The three-way catalytic converter differs in that and also facilitates a reaction that allows for the reduction of nitrous oxide products [33]. Both types use porous clay sheets impregnated with noble metals such as palladium and platinum to facilitate the complete oxidation

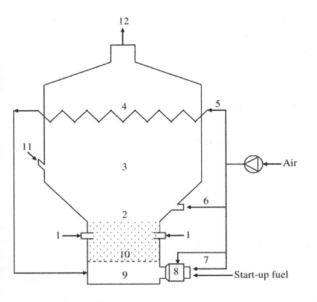

Figure 1.8: Fluidized bed waste incinerator diagram.

of combustion products. An advantage of the catalytic converter is that it helps to re-duce atmospheric pollutants like carbon monoxide and increase the flue temperature, which allows for more stable PCDD/F formation [34]. Kaivosoja et al. [35], however, found that the catalytic converters cause a significant increase in the release of PCDDs/Fs and chlorophenols, both beyond the limits for MSW in the EU. This release would put extra strain on the filtration system, therefore, limiting the hydrocarbon and particulate matter interactions. The value of noble metals needed to make the catalytic converter must also be taken into consideration as they are expensive and rare [35]. Reburn tubes present a potential alternative.

Reburn tubes or secondary burn tubes use heated oxygen to facilitate the burn of smoke products. There are multiple ways to force this extra oxygen. Tubes can be added to the top of the apparatus that allow for increased air flow, or a secondary air source can be added to the back of the burn allowing for air flow to occur over the initial burn. These are cheap and can be executed with pipes found at the e-waste site. They result in a reduction of CO and particulate emissions – the same as the cata-lytic converter – with much less economic and manufacturing constraints.

1.2.4.4 Automatic air intake regulators

Air intake regulators are regularly seen on wood burning stoves. These regulators can be used to adjust the fire temperature, the flue temperature, and the presence of sec-ondary burn mechanisms. The method for air intake regulation can be as simple as a flap, which raises or lowers to allow for air to enter or can be as complex as Wi-Fi-enabled chimney and intake controls. These are both examples of tools that could be classified as part of smart development. In our case, the method that seems the simplest is the most ingenious [36]. In this product, the ideal temperature can be set that initially moves the flap to an open position. After the chamber gets hot, the regulation occurs as a temperature-dependent spring expands within the mechanism and forces the flap down over the intake valve to keep the temperature steady. As the burn chamber then begins to cool, the spring contracts causing the air intake to reopen and raise the temperature.

1.2.4.5 Vent pipes

Before installing a burning device, it is imperative to read about what venting pipes are necessary for the application. The type of burning done in the burn box will be closest to that of a wood stove. The burning of wires can reach around 800 °C, meaning the vent pipe must manage hot temperatures, which are primarily only wood stove vents. Class A chimney pipes are used for wood stoves and are often referred to as insulated chimney pipes. The vents are either composed of double or triple walls, but double-walled pipes are the most common. The inner pipe holds the emitted heat and particles from the combustion. The space between the inner and outer pipe functions as an insulator to prevent the outer pipe from reaching too high of temperatures [37]. While this would be a safe option for the burn box design, class A chimney pipes are used because they run through walls and ceilings. A class A chimney pipe is necessary for the burn box if the burning were happening inside, or if it did not meet the necessary clearances from combustible materials. The burn box is going to be outside and far enough from combustible materials to allow us to use a more cost-effective option, a stove pipe. Stove pipes are typically single-walled and require an 18-in clearance from combustible materials [38]. For an 8-in diameter vent that is 48 in long, a class A chimney pipe costs around $360 while a stove pipe of the same dimensions can be < $70 [39]. Stove pipes are not only more cost-efficient but also more commonly found than a class A chimney pipe, which makes it the best option when using recycled materials.

1.2.4.6 Scrubber design

The toxins that do not completely combust within the incinerator portion of our design will need to be caught by the scrubber. The scrubber for the burn box will remove toxic emissions from the wire burning process. The Merriam-Webster dictionary describes a scrubber as "an apparatus for removing impurities especially from gases." The two main types of scrubbers are wet and dry. Wet scrubbers use a type of liquid to dissolve the toxic fly ash from the incinerator. In our case, and in many others, this is water [40]. However, dry scrubbers use a solid matrix to capture the molecules of interest, which usually includes some form of powder. Dry scrubbers often induce some sort of chemical manipulation such as reduction agents and desulfurization [41]. Scrubbers are often implemented in industrial processes requiring smokestacks.

Copper smelting, hospital waste, and MSW burning are three processes that have similar emission profiles to those seen from wire burning sites [21]. Each commonly uses a filter that includes an initial screening component such as a fabric filter, a section to actively exclude certain molecules, such as an activated charcoal filter, and a last section to neutralize any molecules getting through the initial exclusion such as wet deacidification.

The fuel source must be considered when designing a filtration system. As tires and refrigerator insulation are often used for fuel in open burns, we cannot assume that these will be forgone with the implementation of our burn box [3]. Therefore, the toxins released in the burning of these fuels need to be addressed. The main concern here will be the organic toxins and HCN released in these burns [30, 32]. The organic compounds should be captured the same as the DRCs already accounted for; however, the HCN will require special treatment. This should be neutralized through wet deacidification and deacidification [21].

Using materials found in Ghana, we seek to simulate similar filtration systems. A combination of coconut husks and shells could create a layered activated charcoal filter, which allows for size exclusion purification of the smoke [42]. Calcination reactions using heated eggshells can nullify the effects of the toxic heavy metals and acidic compounds. HCl and HCN can also be neutralized by adding a weak base/buffer to the filtration system. Filter life, waste disposal, and specifications will be examined as we continue in our research [43].

Research, however important to the process of design, does not make up for years of experience working in and around a problem. While the designs described above are successful toward their target audience and application, our project's design must differ in its aim for use in the Agbogbloshie e-waste site. The final incinerator and scrubber were co-designed with e-waste workers and students in Ghana to achieve the needs and constraints involved in improving the wire burning conditions. Collaboration throughout the project will hopefully provide a long-lasting, sustainable solution to be developed and maintained independently by the Agbogbloshie e-waste workers.

1.2.5 Co-design process

When working to create a device that will benefit another community, using a co-design process helps to establish independence and self-sufficiency. To accomplish this, the design needs to begin with the community rather than the market. This case study followed a co-design process to ensure that the burn box maximizes its benefits for the end users, the recyclers. The team took an iterative design approach that allowed them to receive feedback on various stages of the design while working with the community. First, they co-designed an initial model with Academic City University College (ACUC) students and then shared it with the recyclers of Agbogbloshie to get feedback. Through each design iteration, the team at Worcester Polytechnic Institute (WPI) and ACUC students received feedback and adjusted accordingly. This co-design process was implemented in hopes that the burn box created would provide the community in Agbogbloshie long-term sustainability by utilizing, maintaining, and developing the technology.

1.3 Methods and materials (the project)

To begin the design work, we first established an understanding that contextualized the problem at hand. Our literature review allowed us to gain a surface-level understanding of the recycling and burn process. Watching documentaries created about Agbogbloshie and visiting the Wachusett Watershed Recycling Center in West Boylston, Massachusetts, in December 2021 allowed us to gain a better understanding of e-waste. After completing our research, we spoke with our co-designers to understand how they recycle e-waste at the site and to gain their input and help in designing our burn box. This section explores our design and research process. We detail our approach toward co-design, and then move to testing and proof of concept.

1.3.1 Co-design with an iterative development process

Through the burn box project, we collaborated with our partners, the e-recyclers, and students at ACUC in Ghana. We had conversations with our two partners to better understand their values. Our co-design process also included weekly meetings with our student partners to review design ideas and converse about what materials and methods should be used to create the burn box.

1.3.1.1 Co-designing with recyclers

We opened discussions with wire burners in Agbogbloshie about their daily activities and the current steps involved with obtaining copper. Within these talks, the health issues involved in the wire burning were discussed, which allowed us to brainstorm safety precautions with our partners. Gathering information from the recyclers at Agbogbloshie gave our team insight into how to create a sustainable design. We learned about the materials available at Agbogbloshie and the size of typical wire bundles burned. This allowed us to establish the material and size parameters for our design. We also learned about the typical number of burns per day, accelerants used to start the burn, ways in which the wire recyclers interact, and safety issues that the recyclers face. This information was the last key piece in establishing the basis for our design.

Once our team had context for starting our designs, we were able to create preliminary designs that we could show the recyclers. We developed two main design ideas based on the information that we received from the recyclers. Our goal with sharing our designs was to receive initial reactions, feedback, and suggestions on each design to give us an idea of the adjustments that needed to be made.

1.3.1.2 Co-designing with Academic City University College students

Our team worked with three ACUC students: Louisa Ayamga, Kwabena Boateng, and Faith Cyril. The students offered a distinct cultural perspective to the design process, which was imperative in the co-designing process. We met with our partners at least once a week to review design ideas and talk about how we can best create a burn box that allows the recyclers to keep their burning process the same while adding in key safety features. The local perspective that the students had allowed our team to see key design elements that we may have been missing or overlooking. As we began looking into building the burn box, the students were also able to see what materials would be best to use based on cost and availability. By co-designing with the ACUC students, our team aimed to create a burn box design that was useful and sustainable for the recyclers.

1.3.2 Building the burn box (prototype development process)

We needed to ensure that our burn box would meet engineering standards and follow the Massachusetts burning guidelines. To address safety concerns surrounding burning on WPI's campus, we chose to use an old wood stove as proof of concept for our burn box design. The stove allowed us to safely perform burns to evaluate all the components in our proposed burn box, namely, our fire bricks and filter.

1.3.2.1 Preparing the wood stove

We went to Higgins Energy Alternatives, which is a fireplace store in Barre, Massachusetts, to conduct research into the operation of wood stoves and obtain used materials. Chris Higgins, Vice President at Higgins Energy Alternatives, let us know that they take customers' old wood stoves when installing replacements. This allowed us to obtain a discarded fireplace insert wood stove to use for our project. The donated wood stove is shown in Figure 1.9A. To prepare the stove for transport, we cleaned out the old ash, sanded, and added a coat of spray paint to cover the rust. This is shown in Figure 1.9B. After cleaning the stove, we talked with some of the workers at Higgins to decide on the best vent pipe for our use. The workers suggested using a 4-ft stove pipe to allow for a strong enough draft to form. When we originally tested the stove at Higgins Energy Alternatives, Ron Higgins, the owner, suggested adding another 2–4 ft of pipe. In total, we added these two pipes and a chimney cap to the height of the stove. The last step to prepare the stove to be transported to our campus was to complete a small burn inside of it to ensure that the stove worked correctly. This was done outside of Higgins Energy, and the stove worked with the stove pipe. Once we knew that the stove was safe to transport, it was moved to the WPI Fire Lab where we could then assemble the full structure.

(A)   (B)

Figure 1.9: Donated wood stove from Higgins Energy Alternatives: (A) wood stove prior to cleaning and (B) wood stove after cleaning.

1.3.2.2 Creating and testing the fire brick

Through the research our team conducted on prior art, we learned about fire bricks. Fire bricks, also known as refractory bricks, can withstand higher temperatures than regular bricks by about 1,037 °C. Regular brick can only withstand temperatures of up to 482 °C, which would not be enough to use for the burn box. After finding out that an individual fire brick costs upward of $3, our team found an alternative solution. We found that you can make your own fire brick using cement, perlite, and sand. To determine if this would be something to use for the final design, our team conducted an experiment.

1.3.2.2.1 Making standard fire bricks

With the help of our professor, Robert Krueger, our team created a wooden form for the bricks (Figure 1.10). The form was made of four sides with no top or bottom so that it could easily be reused before the brick fully dried. The brick form was made to create a brick that was 9 in long × 4.5 in wide × 2.5 in tall.

Figure 1.10: Brick mold.

Once we had our brick form, we could create the brick itself. Based on two recipes for fire bricks we had found on YouTube, our team formed three recipe variations to try, as given in Table 1.2. The two videos we based our formulas on are *How to Make Refractory Fire Bricks for a Forge or Foundry* [44] and *Perlite vs. Vermiculite for DIY Fire Bricks (Comparison)* [45]. Our team calculated the volume of one brick and used that in combination with each ingredient ratio to find the amount in quarts needed for each. Once the dry ingredients were measured into a bucket, we added water while mixing until the desired consistency was reached. For consistency, we were going for "stiff cookie dough" as one of the recipes explained it [46]. We lined the form with plastic and then packed the mix into the form. We let the brick set while we prepared the next mixture. When the next mixture was ready, we would lift the form from the brick, line it with more plastic, and pack it with the next mixture. We repeated this for all three bricks. Once the bricks were made and released from the form, we left them to dry for 2 days before finding they were fully dried.

Table 1.2: Standard fire brick experiment variations.

Ingredients	Brick 1	Brick 2	Brick 3
Perlite	7 parts	7 parts	7 parts
Portland cement	2 parts	4 parts	3 parts
Sand	2 parts	4 parts	3 parts

We performed an initial test with the fire bricks to heat resistance after the bricks had completely dried, approximately 7 days after creation. These bricks were then subjected to a wood-burning fire for around an hour, so they could be raised to temperatures of 500–700 °C and monitored by an infrared thermometer. The bricks were then taken out of the fire and cooled.

1.3.2.2.2 Making rice husk fire bricks

After speaking with our student partners at ACUC, we realized that perlite is not a common material found in Ghana. To make the recipe more accessible, our team looked for alternative materials. We found a research paper about using rice husks to make fire bricks in Nigeria [47]. After speaking with the students at ACUC, we learned that rice husks are more accessible than perlite. Before suggesting using rice husks in the place of perlite, our team had to create our own rice husk fire bricks and assess them. Similar to how our team created the standard fire bricks, we chose five recipe variations by using a combination of the research paper and our standard brick recipe. These are listed in Table 1.3. When making the bricks, our team used the same wooden form that we used when making standard fire bricks. We also followed the same procedure of adding water to the dry ingredients until a "stiff cookie dough" consistency was reached. The form is again lined with plastic, and the mixture is put in the form to set while the subsequent mixture is prepared. The bricks were dried for 5 days before they were heat tested.

Table 1.3: Rice husk fire brick experiment variations.

Ingredients	Brick 4	Brick 5	Brick 6	Brick 7	Brick 8
Rice husks	6 parts	7 parts	6 parts	5 parts	4 parts
Portland cement	2 parts	4 parts	4 parts	4 parts	4 parts
Sand	2 parts	4 parts	4 parts	4 parts	4 parts

Our team heat tested the rice husk fire bricks the same way that we heat tested the standard fire bricks. We placed each brick in a wood fire and left it there for an hour to expose each brick to extended heating.

1.3.2.2.3 Impact testing the fire bricks

To test the strength of the bricks, our team conducted an impact test after the bricks were heat tested. We dropped each of the eight bricks and an industry standard brick from a height of 9.5 ft. The bricks had an impact force between 4.48 and 18.87 N depending on the weight of the bricks. That corresponds to a range between 1 and 4.24 pound-force. The lightest brick was brick 1 which weighed 457 g and the heaviest brick was brick 8, which weighed 1,926 g. The heavier the brick, the greater the impact force. We knew this was true because of Newton's second law of motion. This law "defines a force to be equal to change in momentum (mass times velocity) per change in time" [48]. This means that the force of an object is equal to its mass times its acceleration. When an object is in free fall, as it was in the impact test, the acceleration of the object is equal to the acceleration due to gravity, which is 9.8 m/s^2. This tells us that only the mass of the object influenced the impact force of the brick.

A standard brick is impact tested from 1 m above the ground, which is close to three times less than the height we tested our bricks from. Our team chose such a high height because we wanted to ensure that the bricks were stronger than average. Because our team would be recommending our partners in Ghana to use these, we needed to ensure that the bricks we suggested were structurally sound.

1.3.3 Scrubber development and design

Our main goal was to build and evaluate the functionality of a filter made with materials that could be found at Agbogbloshie. Cell culture-grade powdered activated charcoal was obtained from Sigma Aldrich, and granulated activated charcoal derived from coconut husks was obtained from PureT USA. Calcinated eggshells were created using 15 eggshells. These were washed with deionized (DI) water, dried for a week at room temperature, and then ground in a mortar and pestle. This powder was then subjected to 800 °C for 3 h and left to cool. These materials were used to create model filtration systems for testing, which were scaled up for the prototype.

1.3.3.1 Creation of filtration systems

Our first prototype filter was created using a mixture of calcined eggshells and powdered activated charcoal. A barrier was fashioned using a cotton ball and a small layer of salt inserted to the bottom of a 250 mL column. Then, 7.114 g of calcined eggshell was added to 10.46 g of finely ground activated charcoal in an Erlenmeyer flask. This mixture was added to the column, and then 300 mL of DI water was slowly applied to the top of the column. Pressurized air was then used to propel the leftover water through the column.

The next filtration system was made using a system that separated out the calcinated eggshells and used both types of granulated activated charcoal. A cotton ball was used to plug the end of the column. This plug was then followed by a 3 g layer of coarsely ground calcinated eggshell, and then followed by a 3 g layer of finely ground calcinated eggshell. A 6 g layer of granulated activated charcoal was added on top of the eggshell layers to complete the filtration system.

1.3.3.2 Tests of the filtration system

The two filters were tested using different sized indicator molecules to determine the size range that is best filtered. The indicator molecules and their corresponding pollutants are listed in Table 1.4. This table shows a direct comparison between the molecules released in the wire burning process and the indicator molecules used to test the filter. All data were found using PubChem and the National Library of Medicine [49].

Table 1.4: Comparison of toxin and indicator molecules.

Pollutant	Molecular mass (g/mol)	Indicator molecule	Molecular mass (g/mol)
Polychlorinated dibenzodioxins/furans (PCDDs/Fs)	Min: 218.63 Max: 459.7	Methyl orange	327.33
Polybrominated dibenzodioxins/furans (PBDDs/Fs)	Min: 247.09 Max: 420.88		
Polychlorinated biphenyls (PCBs)	Min: 268.72 Max: 360.9		
HCl	36.46	Methanol	32.04
HCN	27.025	Carbon monoxide	28.01
Cu	63.550	Copper(II) nitrate	63.55*
Zn	65.400	Nickel(II) chloride	58.64*
Pb	207.000	Acetyl naphthalene	170.21

*The molar mass of the dissociated metal ion species was taken.

A 5-mL sample of each indicator molecule highlighted in blue in Table 1.4 was added to a round bottomed flask that was then subjected to its boiling point. The gaseous molecule was run through the column until almost no sample remained. The headspace above the filter was then sampled and gas chromatography (GC) was run to determine the presence of each indicator molecule.

For the other larger molecules, this same method could not be achieved as their boiling points were too high. To understand how the filter works with these molecules, we used a gravimetric approach. We added 5 mL of a 0.5 M solution of each molecule to the top of the column. The column was then subjected to a high-pressurized air system to force the sample through the filtration system. The overall concentration of the filtrate was then ascertained using a standard curve.

1.3.3.3 Translation to the prototype filter

Based on the most successful version of filter design as previously tested, the same amounts of each filtration layer were then translated to the filtration system created for the prototype. A large empty can was used as the housing to construct the prototype filtration system. Each layer was constructed by wrapping lesser amounts of eggshell and activated charcoal in a wire mesh.

1.3.4 Prototype testing

After the components of our prototype were tested separately, we joined them together for a conclusive proof-of-concept test. We first assembled the burn box components together by adding our fire bricks inside the wood stove and fitting the filter into the chimney. We then proceeded to execute a wood burn to ensure that with all the new components the wood stove was still able to function properly. After this we proceeded to conduct a wire burn following the process we learned from the wire burners.

1.4 Data and results

This section details the findings of our different experiments. We initially conversed with the wire burners to gain a better understanding of their wire burn process. We then began executing our own experiments, eventually creating two unique design concepts. These required the development and testing of heat-resistant fire brick and a filtration system. These experiments culminated in our final project prototype, which was subsequently tested.

1.4.1 Preliminary conversations with our partners in Ghana

We began with a conversation among Julian Bennett, a student at Academic City University College (ACUC) in Ghana, and Adam Suale, a worker at the e-waste site. This provided insight into a few materials that may be helpful in our designs, which included computers, air conditioners, old cars, and engine parts. From there, we spoke with wire burners, Mohamed Awal and Godfred Abeerengya. Mohamed and Godfred work as a team when they burn wires and discussed their process with us, which is shown in Figure 1.11. Together they can process about 20 bundles of wire, earning up to Cedi 140 or US $20 a day. This was not the life that they chose, but it is what they do to survive and make a living. Godfred has burn scars up his arms from working with the wires. We were asked about and thanked for the work we are doing to make this process safer for everyone involved.

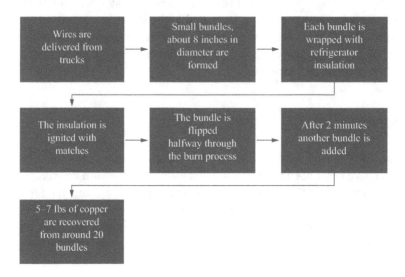

Figure 1.11: The burning process.

From this meeting, we could narrow down the design components necessary for the burn box. Safety was the number one concern; we need the box to insulate from both the heat and smoke. We also learned the importance of flipping the bundle of wires halfway through the burn process to allow the other side to burn. This could be due to the lack of oxygen on the underside of the bundle of wires. We could add airflow with our design; however, mechanically including a component that will allow for the wires to be flipped will allow for a safer experience for the burners, which does not change their current patterns. We also needed to consider the use of fuel for our design as we were told of two new fuel sources, air-conditioner insulation and a liquid fuel (discussed as being petrol or kerosene), used for larger wires or when it is raining. Finally, we also learned a box that opened as if it were an oven was desired over a top-loading box.

1.4.2 Burn box design development

The burn box design was created and developed on SolidWorks with the intention of blueprinting a detailed schematic that could be referred to in the future. The WPI student team collaborated with ACUC students to modify and add components that could improve the performance of the device and be readily available in Ghana. The final SolidWorks file now could be tested through thermal simulation to determine the viability and safety of the design.

1.4.2.1 Initial design ideas and iterations

Before meeting with the ACUC students, our team brainstormed some initial designs to be further developed into a final product. These two designs focused on overcoming the limiting parameter of the heat created through wire burning. Through our team's background research, the material of the burn box was required to handle temperatures up to 800 °C. The design in Figure 1.12 is mostly composed of fire brick and mortar, with only the door, air intake, and chimney being stainless steel. The design in Figure 1.12B consists of a 50-gallon, stainless-steel barrel frame with a handle, air intake, and chimney additions also made from stainless steel. This burn box design was then placed on a fire brick patio, which can allow for multiple barrels to be used on a large area patio simultaneously.

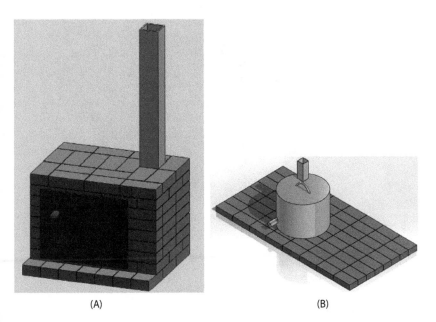

(A)   (B)

Figure 1.12: Initial burn box design ideas: (A) brick design and (B) steel drum design.

Our group presented both designs to the ACUC students in our first weekly meeting. Both groups collaborated on the design iterations, researched materials, and components that could be used in the burn box. We developed a burn box design that incorporated components from both the brick and steel drum designs in Figure 1.12. This design, shown in Figure 1.13, includes the same frame as the brick design but adds the steel drum on the inside of the brick, a grate, and an air intake. The steel drum was added to improve the structural strength of the brick around the frame. The grate was included to prevent a buildup of ash and past burn residue from being in contact with future wire burns that could cause a greater number of hazardous emissions to be produced.

In addition to major design changes, the ACUC students informed us that most brick molds used in Ghana are British standard size (215 mm × 102.5 mm × 65 mm). All the bricks in the burn box design iteration, shown in Figure 1.13, were then changed to British standard size to simplify the brickmaking process.

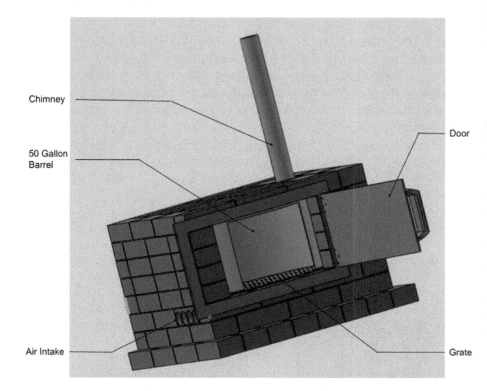

Figure 1.13: Burn box second design iteration.

1.4.2.2 Final design and simulation

The final design for the burn box, shown in Figures 1.14–1.16, added multiple components, and resized to accomplish the device's goal. The chimney was changed to be 8 ft tall with an 8-in outer diameter and a 7¾-in inner diameter. This alteration allows for greater airflow and stack effect to manage the considerable amounts of smoke created from wire burning. An ashtray was added to safely and easily dispose of the ash and other residue created from the burning. The width of the door was increased to cover the whole front of the brick formation, allowing for full access to the inner chamber. The steel drum was removed to make space within the burn box and was instead replaced with thin metal beams to support the brick ceiling.

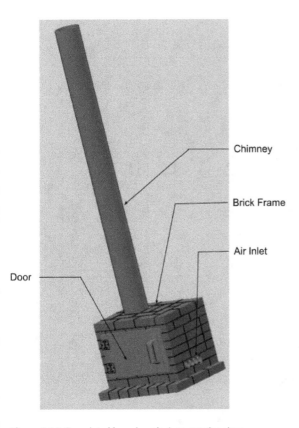

Figure 1.14: Completed burn box design: exterior view.

SolidWorks granted our team the ability to create engineering drawings and simulate thermal loads on the system, shown in Appendix E. By having the actual geometry and size of our design in SolidWorks, we could find the temperature of different components when a thermal load is applied. The chimney, grate, and door were made of stainless steel with a thermal conductivity value preset of 37 W/(m K). The brick and mortar were given a thermal conductivity value of 2 W/(m K) (*Thermal Properties of Cement Mortar*, n.d.). The simulation was set up so that the 8" × 8" spherical ball on the grate, acting as a bundle of wires, was set to 800 °C. This sphere was set to 800 °C to imitate the maximum temperature of the wires during the burn. The sphere was set as copper with the preset thermal conductivity value of 390 W/(m K). The heat from the sphere was conducted to the grate and all other components in contact. Convection values on all outside surfaces from the outside air was given the temperature of 22 °C and a 25 W/(m K) convection coefficient. Temperatures of 415 °C for the inner chimney, 100 °C for the inner brick walls, and 200 °C for the inside of the door were given from prior heat test data to evaluate the resulting outer temperatures of the burn box. The outside of the chimney temperature decreased slightly from the inside,

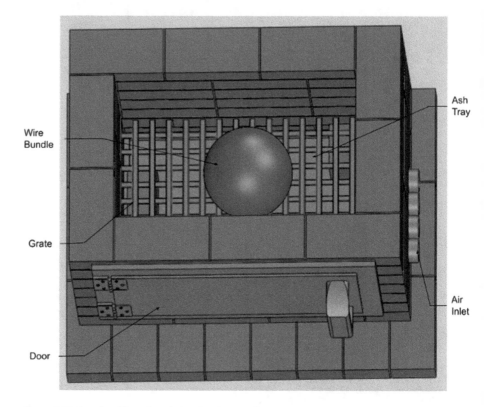

Figure 1.15: Completed burn box design: interior top view.

415 to 413 °C. The outside of the door temperature decreased from the inside, 200 to 186 °C. The outer brick wall temperature decreased from the inside, 100 to 51 °C. This thermal simulation and list of temperatures provide another approach to verify that the materials will not reach above their temperature limits and that the burn box will be safe to operate.

1.4.3 Fire brick experiment

To determine the best mixture for both the standard fire bricks and the rice husk fire bricks, our team heat tested and stress tested each brick. This allowed us to ensure that the bricks will withstand the hot temperatures in the burn box, as well as create a strong structure.

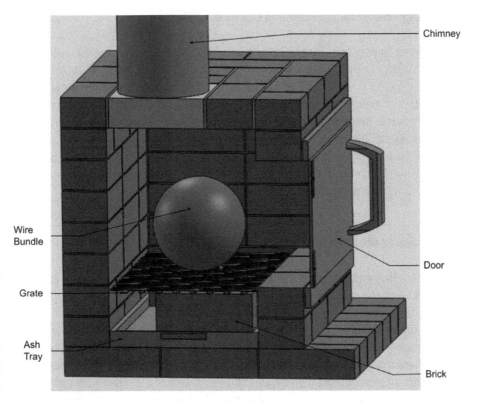

Figure 1.16: Completed burn box design: interior side view.

1.4.3.1 Standard brick heat test results

The three standard bricks were heated in a wood fire for 1 h. The fire reached upward of 800 °C. Each brick was then removed from the fire and given time to cool. None of the bricks appeared to have lost structural integrity during the burn.

1.4.3.2 Rice husk brick heat test results

Each brick was heated in a wood fire for about an hour to see how they held up to heat. Before burning the bricks, it was visible that brick 4 seemed to be the least sturdy. Once the bricks had been in the fire for an hour, we removed them for cooling. Brick 1 began to break as it was removed from the fire; brick 5 had a small corner missing from it; and bricks 6, 7, and 8 were fully intact. This followed our initial assumptions as bricks 3, 4, and 5 had the most cement and the least amount of rice husks. The last test to be performed on the bricks to determine the best mixture was a stress test.

1.4.3.3 Impact test for standard and rice husk bricks

Through our impact test, our team saw that some of our fire bricks could withstand a greater impact force than an industry standard fire brick. After each brick was dropped from 9.5 ft, it was inspected. Bricks 1, 2, 4, 5, and the industry standard brick all failed. This meant that they broke into more than two pieces. Bricks 3 and 6 had partial failures. Brick 3 broke into half, but the break was clean and there was no other damage to the brick. Brick 6 lost pieces from three of its corners, but most of the brick was still intact. Bricks 7 and 8 were the only two bricks that were dropped and had no change in their structure. Images of the industry standard brick and bricks 3, 6, 7, and 8 post-impact test are shown in Appendix D.

1.4.3.4 Overall fire brick results

Throughout the tests, our team was able to identify the best mixtures of both standard fire bricks and rice husk fire bricks. One of the main factors our team identified that was imperative for having a structurally sound fire brick was the amount of cement used in the mixture. The more the cement used, the better the mixture combined. Having more cement also increased the weight of the brick. To compare all the bricks that our team made, we made a table showing the visual observation of each brick, along with the results from the heat test and the impact test. Brick 2 with mortar was not impact tested because we were evaluating the heat capacity of the mortar as opposed to heat and strength of the brick. This is given in Table 1.5.

Table 1.5: Fire brick test results.

Brick	Visual observation before tests	Withstand 800 °C	9.5 ft drop test
Industry standard	Fully intact	Pass	Fail
Brick 1	Fully intact	Pass	Fail
Brick 2	Fully intact	Pass	Fail
Brick 3	Fully intact	Pass	Partial fail (broke in half)
Brick 4	Structurally unsound (easily crumbled)	Fail	Fail
Brick 5	Fully intact	Fail	Fail
Brick 6	Fully intact	Pass	Partial fail (partial break)
Brick 7	Fully intact	Pass	Pass
Brick 8	Fully intact	Pass	Pass
Brick 2 with mortar	Fully intact	Pass	Not tested

After reviewing the results from the heat test and the impact test of the bricks, our team decided to recommend the mixture from Brick 7 (Figure 1.17). We chose this brick because it withstood the high heat temperatures, did not fail during the impact test, and contains more rice husks than brick 8. Having a greater percentage of rice husks is beneficial because it is a material that we believe will be the least expensive and most available material in the mixture. Through our tests we also found that mortar made from three and a half parts sand to one part cement has the capacity to withstand the high stove temperatures without compromising its properties. Our team recommends the use of this mortar in conjunction with brick 7 when constructing the burn box.

Figure 1.17: Brick 7.

1.4.4 Filter tests

The first filter iteration did not allow for DI water to be passed through the activated charcoal. This meant that not only would this prevent the creation of a draft for the burn box chimney but also did not allow for the washing of the activated charcoal that clears the pores within the charcoal granules. Due to these issues, the second filter iteration, which is shown in Figure 1.18, was used for testing.

The tests used to determine the filter's ability to absorb HCl and HCN showed that these molecules would escape. The indicator molecules bubbled through the filter, methanol, and carbon monoxide were detected in the GC tests of the headspace above the filter after each test. As these toxins are similar in size to the nitrogen, oxygen, carbon dioxide, and water vapor species found in air, they needed to go through the filter, or we could not create a draft. To combat this process, we added calcinated eggshells to the filter design, allowing for the greater absorbance of the acidic species being released.

The larger species were found to be absorbed with differing success. Copper and zinc constituting the lower end of the molar masses that we hoped to replicate through our experiments showed that they were only absorbed to a small extent within the filter. The copper salt assay in fact resulted in an eluent hat that was cloudy and therefore

Figure 1.18: Filter created for lab testing.

could not be spectroscopically determined; however, upon visual analysis, the color was slightly lighter than the original sample applied to the column. Nickel salt, which was used to simulate the size of zinc, showed a 0.1 M decrease after being passed through the filter. The data collected for naphthalene and methyl orange, representing the larger polychlorinated hydrocarbon species and lead, showed that the filtrate collected had amounts of each substance that were under the limit of detection. The full data and graphs are given in Appendix C. In Figure 1.19, the results of the two metal tests and the test using acetyl naphthalene are shown, while Figure 1.19A shows results of the two metal tests. The first three test tubes seen from the front of the rack backward are the standards that were assessed and the fourth test tube in the row is the unknown that went through the filter. Figure 1.19B shows the acetyl naphthalene filtration in progress, and some of the acetyl naphthalene can be seen beaded up on the activated charcoal.

The next step in the filter development process was to convert these findings into a filter that could be fit into our 8-in diameter chimney. One main design concern was making sure that enough draft could get through the chimney. To address this, the filter was built with multiple layers, each one composed with the same bed height as evaluated above, however, only covering at most 50% of each layer. This filter was built with four layers in an old tomato can using a ¼ × ⅛ in wire mesh to support each layer composed of approximately 10 g of activated charcoal made from coconut

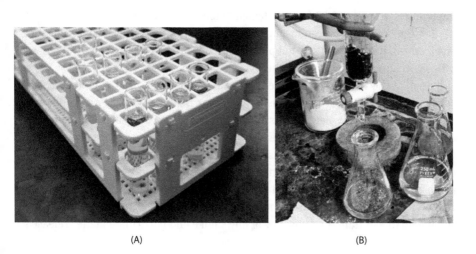

(A)   (B)

Figure 1.19: Filter testing: (A) metal ion tests and (B) acetyl naphthalene test.

husks and 5 g of calcinated eggshell. The resultant filter in Figure 1.20A and B shows a three-quarter view and a top-down view, respectively.

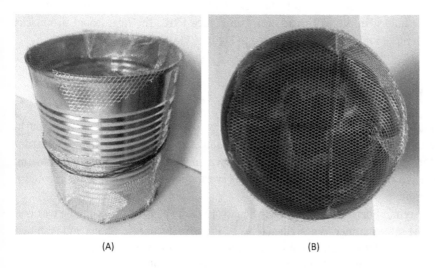

(A)   (B)

Figure 1.20: Prototype filter for testing: (A) side view of filter and (B) top view of filter.

For the creation of this product in Ghana, the activated charcoal can be produced in multiple ways. The testing above was done using coconut husks to create the charcoal through a sustained burning process. This charcoal was then activated through exposure to NaCl to create pores in the charcoal, constituting a slightly less dangerous and more available means of acid activation. However, if a device is available that can maintain temperatures around 800 °C for a sustained amount of time, steam activation is an option that can create a more uniform pore size, but only when the elevated temperature is maintained. If the pore size is not properly established, the filter design will not work as the molecules needing to escape through the filter may not do so as easily.

Based on the amount of activated charcoal and eggshell used to create the filter, if properly maintained, it should last 4–6 months before needing to be replaced. At the end of each week, when work is light for the wire burners, it would be ideal for both the chimney and filter to be cleaned. The chimney should be scrubbed to get rid of any built-up, highly flammable ash, and the filter should be taken out and rinsed gently or disturbed to expose the inner charcoal granules and their pores so that they can absorb the molecules presented to them. The filter was fitted into the top of our design for easy removal and maintenance, so this process will require a ladder, but it should help to maintain the buoyancy of any smoke and particulate matter released.

1.4.5 Proof of concept

Our final proof of concept was built out of the wood stove, a chimney cap, and two chimney pipes obtained from Higgins Energy Alternatives. The filter was placed within the chimney cap using a metal ring to block air flow, a wire mesh, and three screws to hold it in place. This is shown in Figure 1.21C. The heat-tested fire bricks were also placed in the stove as pictured in Figure 1.21A and B.

1.4.5.1 Final burn test

Our final burn test was conducted in two parts: the wood burn and the wire burn. The wood burn lasted 8.5 min after full ignition. The temperature data taken from both the chimney and wood stove are shown in Figure 1.22. The smoke flowing through the chimney, even when the doors were open, indicated that we created a strong draft.

The second burn test was conducted using a bundle of wires. This bundle was created in the same manner as described to us by the wire burners in Ghana and is shown in Figure 1.24A. After ignition, chimney temperatures were recorded every minute; the graph generated is shown in Figure 1.23. This test saw a small amount of smoke leaking in between the chimneys. As the fire was seen to increase after around 4 min, the chimney temperatures slightly lagged behind. When the chimney began cooling, we saw an increase in smoke both through the top of the chimney and escaping out of the front doors. We then locked the doors to prevent more smoke from escaping. By the 10 min mark, we noticed a diminished amount of smoke and a lighter coloration indicating that most of the plastic had burned. We waited until minute 24 to close the air inlet and then let the fire to burn it out.

(B)

(A)   (C)

Figure 1.21: Fully assembled wood stove burn box: (A) Full view of the stove; (B) view of fire bricks placed in stove; and (C) filter inserted in the chimney cap.

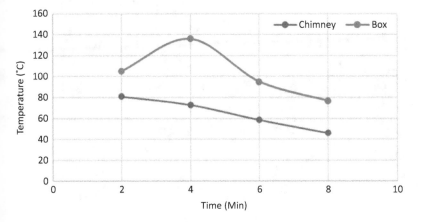

Figure 1.22: Final wood burn temperatures.

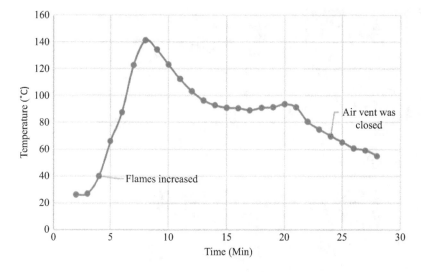

Figure 1.23: Wire burn chimney temperatures.

The day after the burn test, we completed the cleanup and examined the bundle of wires. While there was still a burnt plastic residue on the wires, this was easily broken off by physical manipulation. The resultant wire bundle is shown in Figure 1.24B.

(A)   (B)

Figure 1.24: Wire bundle before and after burn test: (A) wires before being burned and (B) wires after being burned.

Through our proof-of-concept test, our team ensured that the components we tested separately would work together in our final design. We determined that our filter not only works but also allows enough of a draft to be created in the chimney. We also discovered we were reaching temperatures that were high enough to create the draft we needed and to achieve the burn results we wanted. Because of these tests we are recommending a fire brick and mortar that will handle the burning process. Based on the success of this final test, we believe that an unchanged wire burning process will be possible inside of a burn box implementing our design criteria.

1.5 Discussion

Smart villages were originally built to service the Global North, while the rhetoric around smart villages that has changed their implementation has remained stagnant. Previously, smart villages were implemented in lower income countries as a way of control, or "white saviorism." In our project, we sought to challenge these notions. Instead of working "for" our end users, we were collaborating with them. We were able to discuss distinctive design components and processes and understand exactly what was important. Our goal, rather than to revolutionize a process that we know less about, was to learn from our partners and change as little as possible while helping to make their process safer through our designs.

Rather than relying only on our views of what science is, we chose a lab-based synthetic, neat form of inquiry to look toward knowledge bases that were different from our own. We learned from the wire burners and the skills they had learned as a way of surviving, while they may not have measured to the thousandth place the mass of the wires they were burning and the fuel they had knowledge beyond that we could have hoped to obtain in these 8 weeks of study. We relied on their years of experience in creating and modifying our designs to something sustainable and useful. Not only that, but we were also able to better convey our ideas with our partners through these discussions.

Our strengths in design truly came from our partners and their main concerns. No matter who we were speaking with, safety was the number one concern. We found that certain design aspects allowed for safer designs and were more desirable. We needed to collaborate with the community to draft ideas that would work for different burners and protect those affected. Our designs therefore found their strength in those we were working with. Next time, we would contact the mechanical engineering students in Ghana sooner so we could have more initial input before we began concretely conceptualizing our ideas. This would have allowed us to do an initial brainstorm with our partners before showing our ideas, which may have limited or directed their creativity.

Generative justice is defined in different ways. While some may reduce it to simple concepts such as value, generative justice has much more depth. In the end, it is a concept of self-sufficiency in which people maintain ownership of their creations, whatever those may be. In our project, we sought to promote this concept by making sure the wire burners maintained ownership over their process and how they choose to go about their work. Every design step was made with either confirmation or requests in mind from the wire burners or the ACUC students. We were not designing for the wire burners but designing alongside them and that is what generative justice really meant to our team.

1.6 Conclusion

Our team created a proof-of-concept burn box that was able to lay the foundation for a burn box being built in Ghana. While our team used an old wood stove as the structure for the burn box, we formed a design model in SolidWorks that we recommend being followed in Ghana. The design model includes a fire brick structure, a steel door, an 8-ft chimney with support rods and filter, and a burning platform composed of a grate and ashtray. Our team created our own viable fire bricks using a mixture of cement, sand, perlite, plus a mixture of cement, sand, and rice husks. Rice husks are more commonly found in Ghana than perlite, but our team believes that in the future an attempt can be made using coconut husks in the place of rice husks. Coconut husks are the most readily available of the three potential ingredients we listed.

1.6.1 Future recommendations

While our design is a viable solution through our research and experimentation, we found several additional design options that could create a better burn box. Many of these solutions were mentioned earlier in our literature review. We recommend that, in the future, these burn boxes be constructed with a form of secondary burn mechanism, such as catalytic converters and reburn tubes. Reburn tubes especially could benefit from a form of air-intake regulation as these allow for a secondary flow of oxygen into the box. To maintain the draft needed for the chimney to be the only exit for the smoke, we need some form of air-intake regulation to be added. This could be as simple as the intake regulator mentioned earlier that uses a set point and a temperature-sensitive coil as a feedback mechanism. Outside of these mechanical changes, we also recommend the addition of a chimney cap. We used one for testing; however, different types being studied could be a good option for future research. Finally, we recommend subsequent design iterations continue to be designed closely with the community, and the development process to occur with the burners.

1.6.2 Desired lasting effects

Our design was created by working with our partners in Ghana to fit their needs; however, we hope it has a sort of universality that could be beneficial to other e-waste sites. Since open burning of not only e-waste but also MSW is prevalent in so many countries, our design and process could be applied as an example in these cases. Other teams can work with partners not only sharing in our designs but also building on our use of co-design and generative justice. Eventually through this work, we and whoever else takes up the next project will eventually create a better waste incinerator, still there will always be a better solution that we should always keep working toward.

We hope that eventually a long-term study of the effects of our burn box implementation can one day be executed. Our box, when used and functioning as expected, should promote a decrease in the observed incidence of respiratory disease, burns, and cancer in the surrounding areas. For follow-up studies, the monitoring of soil and air pollution would be beneficial to understanding the effects of both the box and the remaining pollution. We also hope that there will be other designs created within the community, by our partners, or by community observers that function to create this same reduction.

Aside from the recycling activities, air pollution in Ghana is contributed to by using biomass fuels in cooking. Biomass fuels include charcoal sources such as charcoal and firewood. Emissions from these fuels are much higher than those seen in cooking with electric or LPG stoves [50]. There have been policies and incentives initiated in Ghana to promote the uptake of LPG stoves; however, this is resisted due to monetary, safety, and cooking concerns [51]. The problem with this strategy is that it was not thought of considering current practices and the opinions of those it was trying to help.

In designing our burn box, we developed fire brick prototypes and an almost ubiquitous filtration system. In using our product with a few changes, cook stoves could be made that allow for the safe burning of traditional fuels. This would drastically decrease air pollution and the associated negative long-term health effects such as respiratory distress, cancer, and low birth weight.

1.6.2.1 Policy changes in the Global North

In completing our project, we had to create a fundamental understanding of e-waste, where it goes and whom it affects. The answer is that the Global North is creating more e-waste than ever but not enacting the necessary policy changes to limit both this creation and its effects in downstream processing. We hope that by compiling this information through our research, together with our efforts, will allow a demonstration for the need for change. Manufacturing practices should be regulated such that at the end of life when and if these products end up at an e-waste site, they will not create health hazards to those collaborating with them. These could be implemented through a decreased use of BFRs, new insulatory polymers, and better insulation. More recycling programs should also exist, assisted by manufacturers for some recycling to occur in the Global North. The economy at these e-waste sites is dependent on the continued exportation of e-waste from the Global North. Projects working with these sites, such as ours are important to co-design solutions to the problems this causes; however, they do not address the underlying issues that created this unbalanced dependency. All we can hope to do is to promote generative justice in our design practice and work to bring light to these issues.

References

[1]   E-waste in Developing Countries: Treasure to Trash? (2020, November 18). *BORGEN*. https://www.borgenmagazine.com/e-waste-developing-countries/

[2]   Turrentine, J. (2020, April 20). At 59 Million Tons, our E-waste Problem is Getting Out of Control. NRDC. Retrieved January 24, 2022, from https://www.nrdc.org/stories/59-million-tons-our-e-waste-problem-getting-out-control.

[3]   Amankwaa, E. F. (2013). Livelihoods in Risk: Exploring Health and Environmental Implications of E-waste Recycling as a Livelihood Strategy in Ghana. *The Journal of Modern African Studies, 51*(4), 551–575. http://www.jstor.org/stable/43302038.

[4]   Zhang, K., Schnoor, J. L., & Zeng, E. Y. (2012). E-waste Recycling: Where Does it Go from Here? *Environmental Science & Technology, 46*(20), 10861–10867. https://doi.org/10.1021/es303166s.

[5]   Liu, Q., Cao, J., Li, K. Q., Miao, X. H., Li, G., Fan, F. Y., & Zhao, Y. C. (2009). Chromosomal Aberrations and DNA Damage in Human Populations Exposed to the Processing of Electronics Waste. *Environmental Science and Pollution Research, 16*(3), 329–338. 10.1007/s11356-008-0087-z.

[6]   Chen, A., Dietrich, K. N., Huo, X., & Ho, S. M. (2011). Developmental Neurotoxicants in E-waste: An Emerging Health Concern. *Environmental Health Perspectives, 119*(4), 431–438. https://doi.org/10.1289/ehp.1002452.

[7]   Hamouda, K. & Adjroudi, R. (2017). Electronic Waste Generation and Management in the Middle East and North Africa (MENA) Region: Algeria as a Case Study. *Environmental Quality Management, 26*(4), 5–16. https://doi.org/10.1002/tqem.21500.

[8]   Kahhat, R., Kim, J., Xu, M., Allenby, B., Williams, E., & Zhang, P. (2008). Exploring E-waste Management Systems in the United States. *Resources, Conservation and Recycling, 52*(7), 955–964. https://doi.org/10.1016/j.resconrec.2008.03.002.

[9]   "The World Has an E-waste Problem." (2019, May 23). *Time Thermal Properties of Cement Mortar*. (n.d.). Materiales de Construcción. Retrieved March 4, 2022, from https://materconstrucc.revistas.csic.es/index.php/materconstrucc/article/view/2293/3054 .

[10]   Orisakwe, O. E., Frazzoli, C., Ilo, C. E., & Oritsemuelebi, B. (2019). Public Health Burden of E-waste in Africa. *Journal of Health and Pollution, 9*(22), 190610. https://doi.org/10.5696/2156-9614-9.22.190610.

[11]   Perkins, D. N., Brune Drisse, M.-N., Nxele, T., & Sly, P. D. (2014). E-waste: A Global Hazard. *Annals of Global Health, 80*(4), 286–295. https://doi.org/10.1016/j.aogh.2014.10.001.

[12]   *E-waste in the EU: Facts and figures (infographic) | News | European Parliament*. (2020, December 23). https://www.europarl.europa.eu/news/en/headlines/society/20201208STO93325/e-waste-in-the-eu-facts-and-figures-infographic

[13]   Robinson, B. H. (2009). E-waste: An Assessment of Global Production and Environmental Impacts. *Science of the Total Environment, 408*(2), 183–191. https://doi.org/10.1016/j.scitotenv.2009.09.044.

[14] *Circular economy action plan*. (n.d.). Retrieved February 10, 2022, from https://ec.europa.eu/environ ment/strategy/circular-economy-action-plan_en

[15] Wienia, M. (2009). *Ominous Calm: Autochthony and Sovereignty in Konkomba/Nanumba Violence and Peace, Ghana*. Leiden: African Studies Centre. Retrieved from https://hdl.handle.net/1887/14508.

[16] Cogut, A. (2016, October). Open burning of waste: A global health disaster. R20 REGIONS OF CLIMATE ACTION. Retrieved December 14, 2022, from https://regions20.org/wp-content/uploads/ 2016/08/OPEN-BURNING-OF-WASTE-A-GLOBAL-HEALTH-DISASTER_R20-Research-Paper_Final_29.05. 2017.pdf

[17] Zhang, M., Fujimori, T., Shiota, K., Li, X., & Takaoka, M. (2021). Formation Pathways of Polychlorinated Dibenzo-p-dioxins and Dibenzofurans from Burning Simulated PVC-coated Cable Wires. *Chemosphere, 264*, 128542. https://doi.org/10.1016/j.chemosphere.2020.128542.

[18] World Bank Group. (2021, June 1). Solid Waste Management. World Bank. Retrieved January 13, 2022, from https://www.worldbank.org/en/topic/urbandevelopment/brief/solid-waste-management

[19] National Center for Biotechnology Information (2022). PubChem Compound Summary for CID 6338, Vinyl chloride. Retrieved January 13, 2022 from https://pubchem.ncbi.nlm.nih.gov/compound/Vinyl-chloride.

[20] Dhoke, P., Bhargava, R. K., & Jain, S. (2014). Assessment of Toxic Potency of Smoke from Burning Materials. *Journal of Basic and Applied Engineering Research, 1*(13), 34–39.

[21] Song, S., Zhou, X., Guo, C., Zhang, H., Zeng, T., Xie, Y., Liu, J., Zhu, C., & Sun, X. (2019). Emission Characteristics of Polychlorinated, Polybrominated and Mixed Polybrominated/Chlorinated Dibenzo-p-dioxins and Dibenzofurans (PCDD/Fs, PBDD/Fs, and PBCDD/FS) from Waste Incineration and Metallurgical Processes in China. *Ecotoxicology and Environmental Safety, 184*, 109608. https://doi.org/ 10.1016/j.ecoenv.2019.109608.

[22] The 12 initial POPs under the Stockholm Convention. Stockholm Convention . (2019). Retrieved January 14, 2022, from http://chm.pops.int/TheConvention/ThePOPs/The12InitialPOPs/tabid/296/De fault.aspx

[23] Costa, L. G. & Giordano, G. (2007). Developmental Neurotoxicity of Polybrominated Diphenylether (PBDE) Flame Retardants. *Neurotoxicology, 28*(6), 1047–1067. https://doi.org/10.1016/j.neuro.2007.08.007.

[24] Chan, G. H. X., Xu, Y., Liang, Y., Chen, L. X., Wu, S. C., Wong, C. K. C., Leung, C. K. M., & Wong, M. H. (2007). *Environmental Science & Technology, 41*(22), 7668–7674. 10.1021/es071492j.

[25] Toms, L. M., Sjödin, A., Harden, F., Hobson, P., Jones, R., Edenfield, E., & Mueller, J. F. (2009). Serum Polybrominated Diphenyl Ether (PBDE) Levels are Higher in Children (2–5 years of age) than in Infants and Adults. *Environmental Health Perspectives, 117*(9), 1461–1465. https://doi.org/10.1289/ehp. 0900596.

[26] *Hydrochloric acid (hydrogen chloride)* . epa.gov. (2016, August). Retrieved January 18, 2022, from https://www.epa.gov/sites/default/files/2016-09/documents/hydrochloric-acid.pdf

[27] Wright, R. O. & Baccarelli, A. (2007). Metals and Neurotoxicology. *The Journal of Nutrition, 13*(12), 2809–2813. https://doi.org/https://doi.org/10.1093/jn/137.12.2809.

[28] Landrigan, P. J. & Todd, A. C. (1994). Lead Poisoning. *The Western Journal of Medicine, 161*(2), 153–159.

[29] Brown, A., Afriyie-Gyawu, E., Rochani, H., & Barham, R. (2017). Academic Public Health Caucus. In Burning of scrap tires, firewood, and liquefied petroleum gas as fuel sources for singeing meat in Ghana: Chemical emissions via smoke and health concerns. Retrieved January 24, 2022, from https://apha.confex.com/apha/2017/meetingapp.cgi/Paper/384422.

[30] Singh, H. & Jain, A. K. (2008). Ignition, Combustion, Toxicity, and Fire Retardancy of Polyurethane Foams: A Comprehensive Review. *Journal of Applied Polymer Science*. https://doi.org/10.1002/app. 29131.

[31] Fisher, D. A., Hales, C. H., Wang, W.-C., Ko, M. K., & Sze, N. D. (1990). Model Calculations of the Relative Effects of CFCs and Their Replacements on Global Warming. *Nature, 344*(6266), 513–516. https://doi.org/10.1038/344513a0.

[32] DeMarini, D. M., Lemieux, P. M., Ryan, J. V., Brooks, L. R., & Williams, R. W. (1994). Mutagenicity and Chemical Analysis of Emissions from the Open Burning of Scrap Rubber Tires. *Environmental Science & Technology, 28*(1), 136–141. https://doi.org/10.1021/es00050a018.

[33] Kašpar, J., Fornasiero, P., & Graziani, M. (1999). Use of CEO2-based Oxides in the Three-way Catalysis. *Catalysis Today, 50*(2), 285–298. https://doi.org/10.1016/s0920-5861(98)00510-0.

[34] Pennise, D. M. & Kamens, R. M. (1996). Atmospheric Behavior of Polychlorinated Dibenzo-p-dioxins and Dibenzofurans and the Effect of Combustion Temperature. *Environmental Science & Technology, 30*(9), 2832–2842. https://doi.org/10.1021/es960112j.

[35] Kaivosoja, T., Virén, A., Tissari, J., Ruuskanen, J., Tarhanen, J., Sippula, O., & Jokiniemi, J. (2012). Effects of a Catalytic Converter on PCDD/F, Chlorophenol and PAH Emissions In Residential Wood Combustion. *Chemosphere, 88*(3), 278–285. https://doi.org/10.1016/j.chemosphere.2012.02.027.

[36] Nederby, L. (2021, February 3). An Automatic Control Device will Save you Time and Money. Aduro. Retrieved March 1, 2022, from https://www.tips.adurofire.com/knowledge-and-useful-advice/light ing-a-fire/automatic-control-device/

[37] *Chimney and Venting Pipe Buying Guide*. (2019, July 9). Fireplaces Direct Blog. https://www.fireplaces direct.com/blog/chimney-and-venting-pipe-buying-guide

[38] Champagne 2019-01-07, C. (n.d.). *The Ultimate Guide To 5 Types of Chimney Pipe: By The Pros*. Retrieved February 21, 2022, from https://www.efireplacestore.com/chimney-pipe-buying-guide.html

[39] *Fireplaces, Stoves, Stove Pipe & Accessories*. (n.d.). Northline Express. Retrieved February 21, 2022, from https://www.northlineexpress.com/

[40] Mussatti, D. & Hemmer, P. (2002, July 15). Section 6 particulate matter controls. US EPA. Retrieved January 20, 2022, from https://www3.epa.gov/ttncatc1/dir1/cs6ch2.pdf.

[41] Sorrels, J. L., Baynham, A., Randall, D. D., & Laxton, R. (2021, April). Section 5 SO and acid gas controls. epa.gov. Retrieved January 20, 2022, from https://www.epa.gov/sites/default/files/2021-05/documents/wet_and_dry_scrubbers_section_5_chapter_1_control_cost_manual_7th_edition.pdf

[42] Cobb, A., Warms, M., Maurer, E. P., & Chiesa, S. (2012). Low-Tech Coconut Shell Activated Charcoal Production. *International Journal for Service Learning in Engineering, 7*(1), 93–104. Retrieved January 18, 2022.

[43] Park, H. J., Jeong, S. W., Yang, J. K., Kim, B. G., & Lee, S. M. (2007). Removal of Heavy Metals using Waste Eggshell. *Journal of Environmental Sciences, 19*(12), 1436–1441. https://doi.org/10.1016/s1001-07420760234-4.

[44] Make It. (2018, March 7). *How To Make Refractory Fire Bricks For A Forge Or Foundry*. https://www.you tube.com/watch?v=2Gs6z–gMQk

[45] Food Related. (2019, October 17). *Perlite vs Vermiculite for DIY fire bricks (Comparison)*. https://www.youtube.com/watch?v=PihOskFJw2s

[46] *How to make refractory concrete step by step 3+ quick to make recipes*. (n.d.). Retrieved February 10, 2022, from https://delftclay.co.nz/how-to-make-refractory-cement-3-recipes/

[47] Ugheoke, B. I., Onche, E. O., Namessan, O. N., Asikpo, G. A., Ugheoke, B. I., Onche, E. O., Namessan, O. N., & Asikpo, G. A. (n.d.). *Property Optimization of Kaolin- Rice Husk Insulating Fire- Bricks*.

[48] *Newton's Laws of Motion*. (n.d.). Glenn Research Center | NASA. Retrieved March 3, 2022, from https://www1.grc.nasa.gov/beginners-guide-to-aeronautics/newtons-laws-of-motion/

[49] U.S. National Library of Medicine. (n.d.). PubChem. National Center for Biotechnology Information. PubChem Compound Database. Retrieved January 26, 2022, from https://pubchem.ncbi.nlm.nih.gov/

[50] Afrane, G. & Ntiamoah, A. (2012). Analysis of the Life-cycle Costs and Environmental Impacts of Cooking Fuels used in Ghana. *Applied Energy, 98*, 301–306. https://doi.org/10.1016/j.apenergy.2012.03.041.

[51] Dalaba, M., Alirigia, R., Mesenbring, E., Coffey, E., Brown, Z., Hannigan, M., Wiedinmyer, C., Oduro, A., & Dickinson, K. L. (2018). Liquified Petroleum Gas (LPG) Supply and Demand for Cooking in Northern Ghana. *EcoHealth, 15*(4), 716–728. https://doi.org/10.1007/s10393-018-1351-4.

[52] Oteng-Ababio, M. (2015). Technology must seek tradition: Re-engineering urban governance for sustainable solid waste management in low-income neighbourhoods. Future directions in municipal solid waste management in Africa. Pretoria, South Africa: African Institute of South Africa.

[53] Oteng-Ababio, M., Amankwaa, E. F., & Chama, M. A. (2014). The Local Contours of Scavenging for E-waste and Higher-valued Constituent Parts in Accra, Ghana. *Habitat International, 43*, 163–171.

[54] Hussmanns, R. (2001). Informal Sector and Informal Employment: Elements of A Conceptual Framework. Paper presented at the Fifth Meeting of the Expert Group on Informal Sector Statistics (Delhi Group), New Delhi, 19-21 September 2001.

[55] Cooke, E., Hague, S., & McKay, A. (2016). The Ghana Poverty and Inequality Report: Using the 6th Ghana Living Standards Survey. *University of Sussex*, 1–43.

[56] Owusu, G., & Oteng-Ababio, M. (2015). Moving Unruly Contemporary Urbanism toward Sustainable Urban Development in Ghana by 2030. *American Behavioral Scientist, 59*(3), 311–327.

[57] Pérez-Belis, V., Bovea, M. D., & Ibáñez-Forés, V. (2015). An In-depth Literature Review of the Waste Electrical and Electronic Equipment Context: Trends and Evolution. *Waste Management & Research, 33*(1), 3–29.

[58] Prakash, S., Manhart, A., Amoyaw-Osei, Y., & Agyekum, O. O. (2010). Socio-economic Assessment and Feasibility Study on Sustainable E-waste Management in Ghana. *Öko-Institut eV in cooperation with Ghana Environmental Protection Agency (EPA) & Green Advocacy Ghana, Ministry of Housing, Spatial Planning and the Environment, VROM-Inspectorate*, 118.

[59] Fujimori, T., Itai, T., Goto, A., Asante, K. A., Otsuka, M., Takahashi, S., & Tanabe, S. (2016). Interplay of Metals and Bromine with Dioxin-Related Compounds Concentrated in E-waste Open Burning Soil from Agbogbloshie in Accra, Ghana. *Environmental Pollution, 209*, 155–163.

Cameron Cronin, Connor Cumming, Aidan Horn, Aaliyah Royer,
Lauryn Whiteside, Osabarima Owusu Baafi Aboagye III,
Barima Opong Kyekyeku, Barima Ahenkora, Barima Osei Adom,
Kwasi Anane Asare, Attah Asante, Hermine Vedogbeton,
Robert Krueger

Chapter 2
Using co-design principles to co-create a regional recycling system in Eastern Ghana

2.1 Introduction

Since society began using plastic, 7.8 billion tons of plastic have been produced globally [1]. This is approximately 1 ton per person. Today, more than 300 million tons of plastic are produced annually [2]. Plastic production is ubiquitous. It is spread throughout industries such as transportation, textiles, and construction. This has led to major global pollution primarily in the oceans and rivers, with more than 8 million tons of plastic ending up in the ocean or major rivers annually [1]. Because waste plastics do not degrade quickly, it takes 20–500 years for most plastics to deteriorate, and the problem of plastic waste is a multigenerational concern.

Low- and middle-income countries (LMICs) are disproportionately affected by the world's plastic waste [1]. First, they import plastic waste from North America and Europe [3]. There are some examples of LMICs responding to their internal plastic waste crises. Policy makers in India, Kenya, and Morocco banned single-use plastics, such as plastic bags and utensils [4]. LMICs, which comprise approximately 84% of the globe's population, resort to improper plastic disposal, including littering, dumping, and plastic burning. Each can cause significant environmental and health issues [5]. Burning

Cameron Cronin, WPI student, Computer Science
Connor Cumming, WPI student, Mechanical Engineering
Aidan Horn, WPI student, Data Science
Aaliyah Royer, WPI student, Industrial Engineering
Lauryn Whiteside, WPI student, Robotics
Osabarima Owusu Baafi Aboagye III, Chief, Akyem Dwenase, Akyem Abuakwa Traditional Area
Barima Opong Kyekyeku, Chief, Akyem Batabi, Akyem Abuakwa Traditional Area
Barima Ahenkora, Barima Osei Adom, Chief, Akyem Tumfa, Akyem Abuakwa Traditional Area
Kwasi Anane Asare, Agricultural Chief, Akyem Dwenase, Akyem Abuakwa Traditional Area
Attah Asante, Education Chief, Akyem Dwenase, Akyem Abuakwa Traditional Area
Hermine Vedogbeton, Social Science and Policy Studies Department
Robert Krueger, Director, Institute of Science and Technology for Development

https://doi.org/10.1515/9783110786231-003

plastics contributes to greenhouse gas emissions and global warming, and exacerbates public health concerns [1].

This chapter presents a case study of our efforts to develop local waste management and recycling initiatives in Ghana's Eastern rural region. In this region's rural communities, inadequate management of plastic waste has resulted in widespread adoption of the harmful practice of burning plastic [2]. A growing awareness of these consequences among the communities fostered a desire to seek alternatives, one being recycling. This chapter maps out a network of potential partners for a community-owned business that is scalable to other regions of Ghana, and across sub-Saharan Africa. Our partners, the chiefs, and their communities collaborated with us to expand an existing plastic collection process to a regional collaborative recycling system. The project, if successful, will provide jobs, educational opportunities, entrepreneurship opportunities, and an adequate alternative to burning plastic waste.

Here we will serve to contextualize the problem of plastic waste management in a global context and outline the co-design framework that governed the design process of the proposed system. This chapter unfolds as follows.

2.2 The problem of plastic waste

Globally waste management, specifically plastic waste, has been a growing issue for both the environment and health of people around the world [6]. This is due to the increased production and use of plastic as well as improper disposal methods [6]. The following sections discuss, albeit not exhaustively, the impacts of plastic waste on the environment and human health.

2.2.1 Environmental impacts of plastic waste

Plastic is made from polymers that consist of petroleum, making it an enduring problem because it does not easily decompose in the waste stream or otherwise. Plastic may break down, but it then becomes another problem: microplastics, which enters the ground, groundwater, and seawater, is capable of bioaccumulation [7]. When plastic is burnt, toxic elements enter the atmosphere, soil, and water. Through the Earth's connected ecosystems, there are multiple pathways of exposure for humans.

Burning plastics poses another issue because it releases dangerous chemicals into the air, which are known to be human health hazards. Plastic is manufactured by chemical materials known as monomers, oligomers, and catalysts. When burning these materials in the open, there is a degradation process where by-products are formed, such as brominated flame retardants, phthalates, potentially toxic elements, dioxins and related compounds, bisphenol A, particulate matter, and polycyclic aromatic hydrocarbons. These emissions are extremely dangerous.

Phthalates and plasticizers in PVC readily bond with fats, allowing the phthalates to be absorbed in the bloodstream easily [7]. This process in the human body impacts metabolites affecting the endocrine system, metabolism, and thyroid hormones. Dioxins are harmful to human health as the half-life for dioxins is 7–11 years, leading to short-term conditions in the skin and long-term conditions such as cancer, reproductive issues, neurological, and immunological disruptions in development. The long-term effects of dioxins are more prominent in children as disruptions occur during development. This, along with other emissions present, aggravates the respiratory system and increases risk of heart disease. Open burning of waste causes an estimate of 270,000 premature deaths [8]. The health conditions that occur from open burning plastic are dangerous and need to be addressed. While many are consuming plastic, some are also inhaling toxins, such as phthalates, which come from the burning of plastic. When plastic is not incinerated in a controlled facility, the emissions lead to health problems, such as premature death, increased heart disease risk, allergies, and asthma [9]. The continuation of plastic production is definite for our future; however, through the implantation of plastic disposal projects globally, we can create a sustainable, healthier lifestyle for ourselves and the environment.

2.2.2 Waste management in Ghana

Despite spending 30–50% of their operational budgets on solid waste management, cities in low- and middle-income countries such as Ghana only collect between 50% and 80% of the waste generated [10]. In rural areas, these numbers drop significantly. When waste is disposed of by official services, more than 80% of all municipal solid waste (MSW) makes its way to landfills in Ghana [11]. Landfills are a popular option in Ghana because of their low cost and convenience, according to the Environmental Protection Agency (Ghana) [12]. There are currently 15 landfills operating throughout Ghana.

Of the waste that is accounted for, 13% of it is burned, buried, or dumped in unauthorized places is the way MSW is disposed of 13% of the time [11]. This is approximately 630,000 tons of plastic waste per year that is littered or burned in Ghana. Burying without burning has very few short-term environmental impacts but can become an issue if done in excessive amounts over a prolonged period near water sources. The toxins from the plastics and other materials can leach into the groundwater. Burning plastic contributes to greenhouse gas emissions as well as emitting toxic chemicals that have adverse effects on people and the environment. Dioxins, for example, have both short-term conditions, such as skin irritation, as well as long-term conditions such as cancer, reproductive issues, neurological, and immunological disruptions in development [7]. Unauthorized waste cleanup and management costs the

Ghanian government US $290 million every year to combat the poor sanitation that results in these negative health and environmental impacts [10].

2.2.3 Plastic recycling in sub-Saharan Africa

Most countries in West Africa, including Ghana, are projected to produce at least 1,000–5,000 (10^3) tons of MSW per year by 2025 [13]. Annually, sub-Saharan Africa produces 17 million tons of plastic waste per year, which is forecasted to grow exponentially as urbanization in the region continues to increase. With a population of 1.14 billion, the per capita plastic usage is estimated at 16 kg/year, though this figure fails to highlight the differing degree of plastic waste generated between urban and rural areas. Many high-density population centers in sub-Saharan Africa fail to manage plastic waste, leading to rates as high as 0.8 kg of plastic waste entering the environment per person per day [14, 15].

Plastic is a common waste product in developing countries like Ghana and other African countries [13]. There are distinct types of plastics for different applications. Most common plastics used in consumer products that end up in MSW are polyethylene terephthalate (PET), high-density polyethylene (HDPE), polyvinyl chloride (PVC), polypropylene (PP), and polystyrene (PS) [16]. PET is used to make water and beverage bottles; HDPE is used for shampoo bottles, milk bottles, and freezer bags; LDPE is used to make the ubiquitous plastic bags and food packaging film; PP is used to make bottle caps and plastic bags; PS is used for plastic cups and cutlery; and expanded PS for hot drink cups and protective packaging. Mixed plastic packaging (trays, tubs, and pots). Plastic collected for recycling is first sorted for polymer type, then shredded, washed, melted, and pelletized before being made into new products [16]. The several types of plastic must be sorted because of their composition, which would impact the melting point of the plastic. Paper and glass waste are recycled in similar processes and can often have an unlimited lifetime in terms of ability to be recycled again, unlike plastic [17].

2.2.3.1 Recycling in Ghana

The issues around plastic are now becoming a major problem in many developing countries like Ghana. Nearly 20% of solid waste in Ghana is plastic [10]. This high demand results in 10,000 metric tons of plastic imported annually to make plastic containers, shopping bags, and popular water sachets [18]. Ghana's two major cities, Accra and Kumasi, are estimated to generate 4,000 tons of waste daily, which includes all the plastic used in every consumer product [19, 20]. The rate of waste generation in the country's urban centers is estimated at .47 kg per person per day [21]. A substantial portion of the plastic waste produced here comes from single-use plastics, like the water sachets and other product packaging. Water sachets are 500 mL polyethylene (PE) plastic bags used to package and sell water to individuals [22]. These water sachets are prevalent because it is commonly believed that the water sachets provide cleaner water than the available tap water, although this is not always the case. Water sachet production is under the authority of Ghana's Food and Drugs Board. Some PE can contain toxic chemicals and can be environmentally damaging if left discarded like they often are in storm drains and along the roads in urban and rural areas of Ghana. Water sachets' ease of use leads to difficulties in dissuading public use of these disposable items. Consequently, the widespread improper disposal of PE is environmentally damaging. When discarded into waterways, blockage leads to flooding during storms and increased disease vector problems. Stagnant water blocked by the plastic is a breeding ground for disease-carrying mosquitoes and other illnesses [9].

2.2.3.2 African solutions to the plastic problem in Ghana

Historically, most plastic waste recycling in sub-Saharan Africa has been facilitated by informal waste-pickers [23]. In the absence of formal programs and government oversight of solid waste management, these workers function as key intermediaries in the recycling process, collecting, sorting, and transporting salvageable waste to buyers. Beyond plastic, this informal economy of recovering value from waste extends into every available niche, from upcycling discarded e-waste to the processing of discarded coconut fiber into usable products led by independent entrepreneurs.

Almost all initiatives around recycling or waste management are focused in urban areas with large centers of waste because that is where the highest profit is found. Recycling programs currently range from community-based to governmental programs [11]. There are few government-sponsored programs, like the World Economic Forum's Global Plastic Action Partnership (GPAP), but the Ghanaian government is looking to change this [24]. There are 25 well-established, nonprofit organizations and private plastic recycling or reuse businesses operating in Ghana [25]. These organizations are smaller and have a limited range in which they operate, usually within urban city centers like Accra. Of the 13% of plastic waste generated in Ghana annually, only 4–5% of the recyclable plastic is recovered from the MSW [13].

While plastic pollution continues to rise in Ghana, so too have creative solutions emerged to tackle this problem. A Ghanaian government initiative joined the World Economic Forum's GPAP in 2019 [24]. Nonprofit community lead organizations are mostly based in urban centers and aim to help reduce the burden of plastic waste on the community by providing collection programs. Some examples of nonprofit organizations include the University of Ghana Plastic Recycling Project (UGPRP), Zero Waste Accra initiative, and Green Africa Youth Organization (GAYO) [26]. There are also private recycling businesses such as Nelplast and Mckingtorch Africa that use technological innovation to create multipurpose processes and material that are sold for profit [27]. These businesses and programs have similar goals to collect and dispose of plastic waste from MSW or to collect and repurpose the plastic.

To scale these efforts, they must be formalized in the Ghanaian context. For example, Ghana-based company Nelplast, founded with the goal of converting plastic waste into building material for houses and roads, received funding from a partnership between the national government and the Danish International Development Agency, successfully developed a cost-competitive process for brickmaking that serves as a viable alternative to imported materials. While these businesses often drive innovation and maintain demand for waste, ensuring less results in landfill, as downstream recyclers, they still rely on the waste-picking model to fulfill demand.

2.2.3.3 Putting the pieces together

In the community of Akyem Dwenase, a small farming town in Ghana's Eastern Region in the Kingdom of Akyem Abuakwa, we partnered with Chief Osabarima Owusu Baafi Aboagye, III, other chiefs, and residents, to design a sub-regional recycling system that capitalized on an existing collection initiative. The town placed designated bins to curb the discarding and burning of single-use plastics, which posed health and safety risks to the population. However, these bins eventually began to overflow, lacking a means of recycling the waste. To close the gap in ways to continually recycle the collected plastic, this project brought together partners from Akyem Dwenase, residents, traditional leadership, as well as entrepreneurs and representatives from surrounding chieftaincies, to address the area's recycling capacity area and provide a long-term generative solution.

2.2.4 Generative justice and plastic waste management

Traditional methods of waste disposal, such as discarding and burning, were satisfactory when most solid waste produced in a rural context was organic. In contrast, these methods fail to capitalize on the material value of plastic waste. A system that collects plastic regionally and reuses it seeks to close the loop, connecting the end of stream waste back to the producers. The partners agreed that any model would need to be self-propagating and promote interdependence, rather than independence. In an independent model, each hub village is responsible for servicing the communities in its radius, as well as developing production and buyer arrangements, effectively one of many "single-celled" operations throughout the Eastern Region. The model grows in number, but the size and logistical complexity of each individual recycling system remain similar. In contrast, an interdependent model incorporates each new village into the partnership and seeks to create a larger hub collecting from smaller hubs. This self-similar quality is well understood to be more conducive to growth, referred to as a nested loop, "in which networks of generative cycles are linking social, technical and ecological value circulation at multiple scales, such that they increase the propagation of these sustainable techno-social structures" [28]. Eglash and Garvey demonstrate that interdependence within a system tends to produce better outcomes for those involved and that these systems are stable and self-forming [29]. This supports that an intervillage partnership organized to collectively recycle their plastic will be self-sustaining and mutually profitable. Thus, our primary design consideration was to ensure that the model could be easily built upon and functionally similar at all levels of scale.

As the partners planned out the local recycling initiative for Akyem Dwenase and our partner villages, the aim was to take advantage of a closed-loop model to foster a generative cycle of value circulation from plastic waste. This first required an understanding of the value generators accessible in a local context, shown in Figure 2.1. We identified three categories of value sources to make up a recycling system: the plastic waste stream, organizational capacity, and our recycling partners. Independently, these elements do not form a self-sustaining and self-directed system of value circulation, as evidenced by the logistical shortcomings of the previous collection initiative. The feasibility of the initiative is limited unless a sufficient scale can be achieved, wherein the costs associated with the collection, preprocessing, and transportation of plastic must be balanced by the volume of plastic collected.

Figure 2.1: Components of a recycling system in rural Ghana. A network of value sources that make up the components of a recycling initiative in rural Ghana, categorized by organizational capacity, plastic waste stream, and recycling partners.

The profits of the recycling system will be returned to the community as an investment into increasing the communities recycling capacity, as well as ensuring that the recycling business established with our proposed system can expand their recycling capacity to meet further demand. This community investment has the added benefit of increasing the future value of plastic sold through preprocessing and improved collection. By meeting the buyer needs, investing the partnership will maximize the efficiency of exchange, ensuring the lowest possible amount of value loss between villages and recyclers. We diagram this exchange in Figure 2.2.

While investing in production capacity and collection efficiency are important to maintaining the profitability of this initiative, there are other sources of value that have yet to be mentioned and have remained alienated from this approach. A key aspect of community investment is in education. Local Ghanaian entrepreneurs in the recycling industry indicated that they would not have grown if not for their focus on experimentation with recycled materials, but this expertise came from exposure, which is lacking in rural areas. Education is not only a means of conveying information but also a method of inducing generative cycles of social and political expression. In addition to encouraging students to transform plastic waste into purposeful items of value to them and their communities, students also must be empowered to challenge the behaviors and systemic pressures that initiated the plastic problem in the first place. Through the incorporation of generative techniques for design, our plan will circulate previously alienated value back into the community providing many sources of value to enrich the local economy and residents.

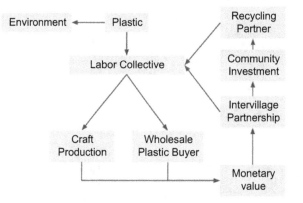

Figure 2.2: Value circulation within an Intervillage Recycling Partnership. Diagram of value flow, diverting plastic material value from environmental sink to community investment through the sale of plastic crafts and wholesale recycling.

Our analysis of the existing recycling initiative reveals that the foundation of a sustainable recycling initiative is present, but that key aspects such as developing partnerships and securing buyers must be implemented to secure the feasibility of this project.

2.3 Designing a new, viable recycling economy

This project was the result of a collaborative effort among numerous partners, stakeholders, and generous consultants across two continents. Indeed, the coauthors of this chapter were the core partners for the co-design of the project. Despite the space that separated us from our partners on the ground, we collaboratively planned, communicated objectives, and examined alternatives. Working around the limitations of distance and language is a critical aspect of co-design in the modern era. In this section, we describe our goals for the design process, and how we achieved those goals.

2.3.1 Three design goals

We entered the design process with three goals to ensure the design is generatively just, culturally centered, and scalable. Generatively just, to achieve meaningful and nonexploitative value exchange between human and ecological sources. Culturally centered, to respect and incorporate existing social relationships, customs, and values. Finally, scalable, to ensure the design can achieve wider adoption while maintaining the two previous goals.

2.3.1.1 Design framework

With goals established, we required a framework for the design process. We centered on the Lingo's (2021), *Empathetic, User-Centered Design Process for Sustainable Outcomes*. Figure 2.3 outlines an iterative process, beginning with an exploration phase where the problem space is better understood through the lens of the local culture. Then we engaged in empathetic inquiry with our stakeholders by consistently iterating our design proposals and redesigning based on their feedback. This allows us to better define the end user needs whereby we can then co-ideate with our end users to better address those needs. Finally, we discuss implementation of a pilot initiative we believe best meets the needs of our partners.

Explore

Community engagement is critical in the co-design process because if collaboration with the local community does not exist, then the co-design process will not occur. The connection with Chief Osabarima began when Professor Robert Krueger, chair of the Social Science and Policy Studies' department at Worcester Polytechnic Institute, introduced our group to him over the phone and we shared our excitement for a potential collaboration to develop a recycling system in Akyem Dwenase. After obtaining input and establishing a connection from Chief Osabarima, we were introduced to

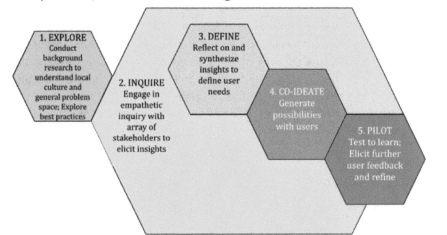

Empathetic, User-Centered Design Process for Sustainable Outcomes

Figure 2.3: *Empathetic, User-Centered Design Process for Sustainable Outcomes.* This diagram outlines the design process to achieve sustainability through empathetic, user-centered design; Once explored, design for a problem necessitates an iterative inquiry stage of redefining and co-ideation before finally piloting the program.

two other important officers, Atta Asante, Deputy Chief of Education and Deputy Chief of Agriculture.

The Deputy Chief of Education was an important part of the co-design process to ensure proper education on the dangers of burning plastic with open fires and the benefits of recycling to engage the community. Buy-in here provided a likelihood where more villagers will take part of the community initiative if the dangers are known. The Deputy Chief of Agriculture was necessary in the co-design process since location is an important part of the recycling system. The Deputy Chief of Agriculture helped determine the locations of open plastic fires that occur in the village. With his input, we determined where to place the recycling bins in the village where many people congregate. This helps is to determine where the plastic will be stored, sanitized, and sorted.

Inquire

The previous recycling program in Dwenase was not as successful as it could be. They were able to provide insight about how the previous system operated by employing children on their weekends to collect the plastic from their bin system, place it into large bags, and store it in the school to be sold for profit in the future. We learned the key hurdle to success was that no one was responsible for emptying the bins. Someone had to take responsibility for this. Furthermore, the representative would have to be someone recognized and respected by the community.

We also inquired about the training for the recycling system. We learned that there was no training in Akyem Dwenase and other villages who were participating in the consortium. Our group had a conversation with Atta Asante, the Deputy Chief of Education for Akyem Dwenase, about how to successfully use the system. With Atta Asante, we created educational infographics to inform the villagers of the importance of recycling and how the system should work. The Chief hosted a recycling workshop to educate the village on how to use the recycling system to be implemented. The educational resources included:

– informing the village of Akyem Dwenase about the dangers of open fire of plastic;
– the purpose of recycling and the profit from the community initiative, which provides an incentive to the villagers to recycle to see that their community will benefit long term and short term by selling plastic and receiving profit as well as creating a centralized makerspace operation in their village; and
– operational features of the new system, including a new flag system to indicate a full bin.

Further inquiry

Akyem Dwenase could not create a recycling business on its own since there is insufficient waste. To make it viable, we engaged chiefs from the nearby towns of Abompe, Batabi, and Tumfa. They showed significant interest and described their own town's system of waste processing systems. We employed lessons from the Dwenase case, and agreed to

– host educational workshops for the villages; and
– identify and delegate a representative for each village to coordinate the plastic collection process and deliver the plastic to the centralized hub.

Our business consultant for plastic recycling has a portfolio that focuses on innovation, sustainability, and community participation. Conversations with the consultant occurred through WhatsApp and Zoom regarding initiating and maintaining a recycling program, with a focus on the process of collection to purchase of plastic. The process begins with citizens picking plastic and cleaning it, where the most valuable plastic is the water sachet, and the common plastic black bags are not commonly utilized. In the case of our recycling system, the plastic would be collected from Abompe, Batabi, Tumfa, and Akyem Dwenase, where the latter would serve as the recycling program's hub. The facilitation in Akyem Dwenase includes collection, sorting, cleaning, and redistribution of processed plastic. We then developed connections with our business partner, planning to expand into Akyem Dwenase. We held conversations over WhatsApp discussing (1) the feasibility and the struggles of the implementation of a recycling system in the Eastern region of Ghana; (2) the need for a larger operation system in Akyem Dwenase to fulfill the business needs; and (3) concerns of our business partner surrounded the start-up and maintaining costs of a recycling system. The highest cost of facilitating the system is the purchasing of machines, electricity, and transportation. Our business partners and the village chiefs agreed to provide some of the financial costs and have local NGOs and foundations cover the remaining costs.

Still more inquiry

Everyone agreed recycling is a promising idea, but there needed to be an incentive to make it happen. Through empathic inquiry we learned that the high price of building materials for housing and other structures and the limited access to off-farm employment opportunities, especially for women and young adults are key issues in the region.

The partners agreed that the waste plastic could be used to develop economic activities and become a source of income and local innovation. The group decided to explore the potential of "MakerSpaces" as incubators for local businesses to develop and innovate with raw plastic and from plastic products. The connections made with businesses can provide plastic products to the community for the villagers to develop new infrastructures such as buildings, bags, and apparel within these MakerSpaces.

Define

Stage 2, Inquire, in Figure 2.3, focused on identifying pain points and desires of our partners, centering on ideals of clean and safe environments, economic empowerment, and concerns over the program as a vector for spreading the COVID-19 virus. As indicated by the diagram, this stage is all-encompassing of the following definition, co-ideation, and piloting stages. Therefore, we worked to ensure that our proposals were not only designed in consideration with the insights gained from our initial inquiry but also that discussions surrounding the proposal always circled back to an assessment of satisfaction and concern with each aspect of the proposal. For example, one proposal iteration advocated for the independent operation of each village in collecting and selling their plastic waste in Accra. This proposal, while intended to ensure less conflict between partners over equitable contribution and compensation, was met with concern from partners that the proposal would not help to develop a strong link between villages. We had failed to identify an equally important objective of our partners: to use the recycling initiative to better strengthen ties between each village, which necessitated a redefinition of our design objective.

After numerous iterations and inquiry, we satisfactorily defined our user needs at increasing scales of involvement. Though many of our discussions were held with village leadership, we still aimed to define the user needs from the bottom up. At minimum, the daily operations of this initiative require labor, capital, and connections. Initial appointed collectors need safe and efficient collection systems and fair compensation for their labor. Thus, they need permanent collection sites that permit sorting and safe, dry storage of plastic waste. In addition, they need transportation to deliver waste to a centralized collection site and potentially expand their operations outside of the village proper. Finally, and most crucially, they need buyers. This has been the largest challenge of this project; however, access to buyers and bargaining power are better facilitated by organized collectives. Thus, we then defined the needs of the organizational structure behind the recycling initiative. To implement a community-wide arraignment, the initiative requires the authority of the chiefs. Luckily, this arrangement is mutually beneficial as many problems facing the chiefs, such as pollution and low employment, are addressed by this recycling initiative. The organizational structure also needs equitable exchange; therefore, there is a need for standardized methods of weighing and compensating for plastic brought to Akyem Dwenase. Moreover, there is a need for regulation and management to ensure that the continued operation of the initiative is accountable to both the laws and customs of the area. At the core of this initiative is the need for the recirculation of value in these rural communities. The need for a clean environment, the need for employment, the need for a fair exchange of value, and the need for a management structure are all symptomatic of a need for each of these components to reverse the drain of value due to the waste of plastic.

Ideate

With these requirements in mind, the team focused on generating possibilities for how to address everyone's needs coherently and consistently. Through meetings with the chiefs from each partner village, we established a baseline for conditions of a successful enterprise in the eyes of our partners to maximize collection, plastic value, efficiency, and opportunity for employment.

Maximizing collection

To address this first concern, maximizing collection, we highlighted the success of the bin system in Akyem Dwenase and provided cost figures for the price of bins and bags to facilitate collection. Plastic consumption data was limited; however, it was noted that plastic usage is often sporadic, correlating with funeral services and other gatherings occurring at the end of each month. These gathering places, churches, marketplaces, and schools are the prime locations for collection bins.

Maximizing value

To address the second point, maximizing value, we needed to address the concern of contamination at the point of collection. Discussions with the chiefs determined the best approach to be a combination of informative graphics on the bins indicating proper use in tandem with a town meeting hosted by a chief appointed educator to convey the necessity of recycling, the requirement of sorting, as well as the community benefits from a successful recycling initiative. Specific buyer relationships needed to be established; thus, a survey team from our partners at Academic City University College would be sent to plastic buyer marketplaces in Accra to gather contact information, wholesale quotes, and plastic demands.

Maximizing efficiency

To address the third point, maximizing efficiency, idea generation centered around how best to transport the largest amount of value with the least amount of cost while ensuring coordination of various operations. One approach focused on establishing independent collection, sorting, processing, and selling operations out of each village. While independent collection and sorting were deemed feasible independently, it was agreed that a portion of profits generated should go toward investments in processing technology centralized in Akyem Dwenase as well as the purchase of one larger transport truck as opposed to a fleet of smaller trucks.

Maximizing employment

Finally, addressing the fourth point, maximizing employment, the discussion focused on where jobs would emerge at each stage of the recycling process. In the initial stage, a chief appointed collector might facilitate local collection to ensure that recycling standards are met, and that collection and transportation run smoothly. As the business grows and capital investments into baling machines are made, additional operators and managers will gain employment as processing increases. As the operation expands, the chiefs acknowledged the role that independent waste collectors would play by collecting from other local towns and granted that a portion of the profits be invested into helping establish these entrepreneurs. Besides direct employment, as profits are collected from the recycling initiative, local projects in need of attention, like building a maternity ward, could be funded, providing additional income to local masons and carpenters.

2.3.2 Community recycling system in rural Ghana

At its core, the recycling system began with the collection and sorting of plastics. Akyem Dwenase and the local villages of Tumfa, Abompe, and Batabi will all be fitted with bin-based recycling systems. We organized a collection and transportation plan using local community representatives for each village. The proposed plan was made up of two major parts: the collection and the profit making. Within each village, the recycling representative will organize the bin collection, plastic storage, and transportation to Akyem Dwenase. The relative location and distance of the primary villages are shown in Figure 2.4. Within Akyem Dwenase, the representative will also have to manage the cleaning and final sorting of all plastics.

We formulated several plans that cover different options for making profit over time. In the next section, we discuss financial costs of the different plans.

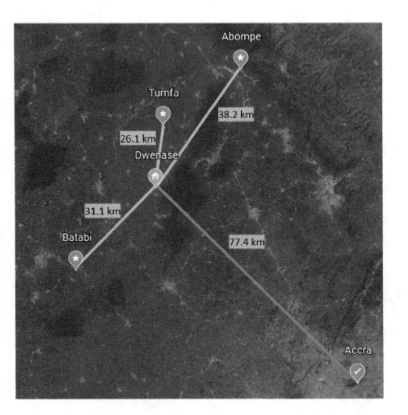

Figure 2.4: Scope of the proposed plan. This map shows the relative distance of the locations we are linking in our proposed collection system.

2.3.3 Collection and transportation system

The team designed the collection and transportation to be modular so that all the villages can operate independently. This allows the community collection system to work best for all the villages and allows for other interested villages to be added to the recycling initiative in the future easily, with low initial investment. The proposed collection system was based on the existing bins that Akyem Dwenase had access to, as shown in Figure 2.4.

The prototype system builds on the initial implementation by adding sorting and a full bin notification system. Sorting will be supported by labeling the bins with images of the types of plastic that can be put in there. The bins will be arranged in pairs where one bin is only for the collection of water sachets and the other bin collects all other types of plastic. The bin pairs, shown in Figure 2.5, will be positioned strategically throughout each village in places that have the most foot traffic and are easy to access.

If a bin is full, the team proposed a flag to be affixed to each bin to indicate how full the bin is (Figure 2.5). This is useful for the community recycling rep to know what bins are close to full so they can be monitored for collection. We also proposed a sliding latch on top of the lid to provide security of the bins. This lid would be made of wood connected with rollers on a track that would be attached to the top of the bin lids shown in Figure 2.5. This lid can be closed and locked when the bin is full or at night if needed. When a bin is full, the community recycling representative can collect the plastic and then reopen the bin. The community recycling representative has other duties such as monitoring the plastic collection and transporting the plastic. The complete process of their roles is shown in Figure 2.6.

Figure 2.5: Existing recycling bins. The existing recycling bins initially provided to Akyem Dwenase.

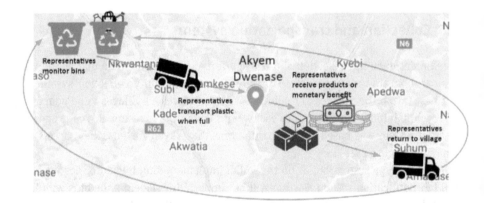

Figure 2.6: Role of community recycling representative. This diagram outlines the major responsibilities the community recycling representative will have. The monitoring of bins is a daily requirement. The transportation of the plastic will begin as a biweekly obligation of the community recycling representative.

2.3.3.1 The system at work

Twice a month, the collection of the plastic will be transported from the other communities to Akyem Dwenase. The route is shown in Figure 2.7. Akyem Dwenase will
serve as the hub for this recycling system. Each local community recycling representative will be responsible for organizing times with the representative in Akyem Dwenase to drop off their plastic. A cross-village panel of management, ABDT (Abompe,
Batabi, Akyem Dwenase, Tumfa) recycling, was developed to oversee all intervillage
operations and the distribution of revenue. This panel can organize different collection times and days based on what works best for each of the communities. In Akyem
Dwenase, there will initially be a dedicated shed to store the collected plastic from all
the villages. Once the plastic is collected and transported to Akyem Dwenase, there
are several operation plans that can be implemented all with different initial investments and profit options.

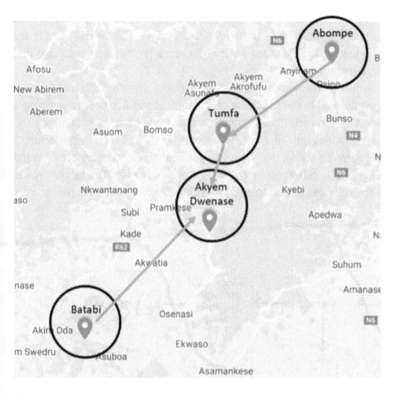

Figure 2.7: Transportation plan. This diagram outlines the travel routes for the surrounding villages to
transport their plastic to Akyem Dwenase. The black circles are the proposed outer areas of the towns to
be included in the recycling initiative in the future. The proposed organization has a combined route to
Akyem Dwenase from Abompe through Tumfa to lessen the overall transportation costs. This route is
43 km long.

Operations

Through the development of the recycling system, we were able to provide several different operation plans, each with its own specific purpose. The main difference in each operation plan is the site of profit generation. We propose three plans: (1) the starting operation; (2) the short-term operation plan; and (3) the local long-term operation. The ABDT recycling panel would be responsible for setting the markers and goals for when or if each next operational plan should be implemented and to what extent. Each plan also has financial variations, which are presented in the next section.

We designed the *starting operation* plan to be the initial approach. This plan requires the least amount of initial investment. The value flow for the profit generation for the initial operation plan is shown in Figure 2.8. This system requires an investment of US $2,000 for bins and bags to set up the collection system in the other three villages, Abompe, Batabi, and Tumfa. This plan requires renting a vehicle to transport the plastic to Accra once a month. The plastic would then be sold to business partners and other plastic recyclers in Accra. The water sachets would be prioritized to be sold to other buyers because our business partner can use the other mixed plastic in their business. The recycling management panel, ABDT recycling, will be responsible for using the profits to pay for the expenses, such as the truck for transportation and fuel cost for intervillage transportation.

A variation on this plan could be realized if the initial investment were raised to US $7,000. This would enable the purchase of a small vehicle for each village and a larger truck. The smaller vehicles, likely motorcycles, would be used to facilitate collection from other locations outside of the main villages as well as transportation to Akyem Dwenase. Figure 2.8 shows that each village could service a small area around them for additional plastic collection by the village representative. The larger vehicle would transport the collected plastic to Accra, helping to lower overall expenses over time.

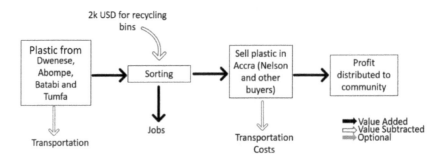

Figure 2.8: Initial plan. This diagram outlines the flow of value for the plastic collected in our proposed recycling system when it is first implemented. The main investment is the bins used to collect the plastic before it is transported to Accra to be sold.

After the goals of the initial plan set by the ABDT recycling panel are met, the next operational plan, short term, begins.

The *short-term operation plan* is remarkably similar to the initial operational plan in terms of the profit generation. Where they differ is that the short-term plan requires a larger investment for equipment. The value flow for the profit generation for the short-term operation plan is shown in Figure 2.9. The investment in the short-term plan would be to purchase a cleaning and baling system for Akyem Dwenase. The price of this system was quoted by our business partner based on the systems implemented in larger scale plastic making businesses in Accra. The cleaning system is used to wash any food or other waste left on the plastic. The baling system is made up of a hydraulic press and wrapping system that can be used to compact more plastic into a smaller volume. Both systems together will add more value to the plastic being sold in Accra. This will bring in more profit per trip as well as provide jobs in Akyem Dwenase for the operation and maintenance of these systems.

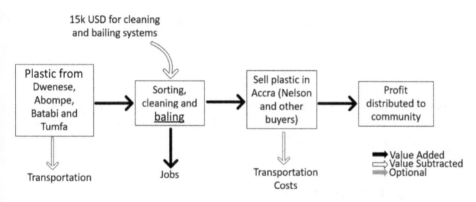

Figure 2.9: Short-term operation plan. This diagram outlines the flow of value for the plastic collected in our proposed recycling system. The main investment is the baling and cleaning system that would be used to add more value to the plastic before it is transported to Accra to be sold.

Lastly, the local long-term operation plan is to set up the operation of a plastic brickmaking business in Akyem Dwenase. This requires an initial investment of US $300,000 to purchase the machines and construct the building in Akyem Dwenase. The value flow for the profit generation for the local long-term operation plan is shown in Figure 2.10. The purchase of land is not necessary as Chief Osabarima Owusu Baafi Aboagye III can provide space for this business to be set up. The cost of the machines is estimated by the larger scale plastic brickmaking business of our business partner in Accra. Once this business is operational, the baling system would no longer be needed, and the larger truck used to transport plastic to Accra can be repurposed to transport the plastic brick produced by this business to the customers in the region. This plan requires a lot of investment but brings lots of value to the community. There will be several jobs available as well as scientific, entrepreneurial, and educational opportunities that can help in the creation of other businesses in the future. This operational plan is the final goal for the proposed recycling system. The bricks made by the business would be used to provide building materials to all the local villages as well as provide more profits to be shared among the villages. The ABDT recycling panel could determine how the profits would be shared as well as decide what next steps this initiative could take in the future. The machines purchased for the plastic brickmaking operation have many uses, so in the future there is the option to create additional smaller businesses to use any leftover plastics.

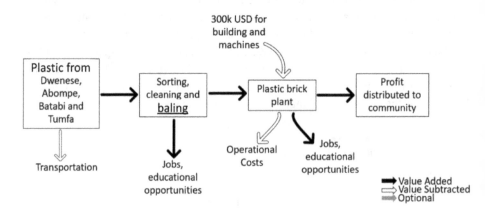

Figure 2.10: Local long-term operation plan. This diagram outlines the flow of value for the plastic collected in our proposed recycling system. The main investment is in setting up a plastic brickmaking company based on the model from our business stakeholder.

2.4 Financial planning

Planning the finances of the recycling initiative led to the creation of three financial models, which were each evaluated to find time to return on investment (ROI), profit after 10 years, rate of return after 10 years, and the ratio of profit to initial investment. To help display calculations, we will walk through the work done to find the outputs for the proposed initial low investment operation plan.

Finding these values began by calculating transportation costs. To do so, we took a vehicle's fuel efficiency and divided it by the fuel cost to give the vehicles fuel price per kilometer. Next, the distances on the primary roads between each village and Akyem Dwenase were found via Google Maps. The estimated distances may be different in practice due to the many unpaved shortcuts that locals may be able to use when transporting between villages. Multiplying the distance between villages and the fuel price per kilometer gave an approximate gas price for each trip. This is reflected in Table 2.1. Note that the gas price to Accra is noted as 150 GHc (US \$21.82). This cost is quoted from local representatives for hiring a heavy truck to take 2 tons of plastic between Akyem Dwenase and Accra.

Table 2.1: Gas cost calculations.

Location	Distance to Akyem Dwenase (km)	Gas price (GHc)
Batabi	50	35
Tumfa + Abompe	62	43.4
Accra	120	150

Next, we calculated the estimated weekly net profit of each route. We multiplied the approximate amount of recycling produced by each village against the current price per kilogram of plastic gave the weekly income of each village. Subtracted from this value was the gas price and estimated weekly expenses. Combining the value of each route gave the overall weekly net profit in Table 2.2.

Table 2.2: Weekly net profit calculations.

Location/ route	Recycled plastic per week (kg)	Price per kilo of plastic (GHc)	Gas price per week (GHc)	Estimated weekly expenses (GHc)	Estimate weekly profits (GHc)
Batabi	306	0.65	35	22.92	140.98
Tumfa + Abompe	994.5	0.65	43.4	22.92	580.105
Dwenase/ Accra	765	0.65	150	22.92	324.33

Combining the estimated weekly profits for all the villages and subtracting weekly expenses gives a net profit of 1045.42 GHc/week. To find time until ROI, the total initial investment of 12,870 GHc (Table 2.10) was divided by the weekly next profit, giving an ROI time of 0.24 years, approximately 3 months.

It must be acknowledged that due to the virtual nature of this project, not all the necessary information was able to be obtained. This lack of material resulted in incomplete calculations and estimated formulas. The data shown below are values that were estimated with the data that was available from our local representatives.

To find the cost of transportation between villages and Accra, an estimation of vehicle fuel efficiency was made (Table 2.3). After looking at the fuel efficiency of popular *tro tro*, models an assumption of 10 km/L was taken.

Table 2.3: Fuel assumptions.

Fuel efficiency (km/L)	10
Fuel cost (GHc/L)	7
Price per km	0.7

Village population and kg of recycled plastic per person per day were needed to estimate how much plastic the villages would produce per week. Village population was estimated through the co-design meetings with the village representatives. Kilograms of plastic recycled per person per week were calculated from data published in [30]. We are operating under the assumption that at peak efficiency, we can collect 90% of the village's discarded plastic. After analyzing the local population within 15 km of each village, we estimated how many more people we could service by implementing motorbikes (Table 2.4).

Table 2.4: Village and recycling assumptions.

Village	Total	Dwenase	Abompe	Tumfa	Batabi
Village pop	**13,500**	5,000	5,000	1,500	2,000
Kg recycle per person per week	**0.170**	0.170	0.170	0.170	0.170
Recycle efficiency goal	**90.00%**	90.00%	90.00%	90.00%	90.00%
Recycle per week initial	**344.25**	127.5	127.5	38.25	51
Additional population serviced from bikes		15%	15%	10%	10%
Updated plastic amount	**2,348.55**	879.75	879.75	252.45	336.6

To calculate profits, one must deduct recurring costs. In this situation, it was unknown how much electricity or building upkeep would cost per week. Instead estimates of $10, $25, and $40 per week were used, increasing with the use of bikes, cleaning systems, and balers (Table 2.5).

Table 2.5: Weekly expense assumptions.

	Price (USD)
Pay	Unknown
Upkeep	Unknown
Electricity	Unknown

We used small-scale versions of the products in our financial models. Without being in Ghana, finding a quote for large machinery is impossible. These prices were taken as reductions of the originals and their outputs were adjusted accordingly (Table 2.6).

Table 2.6: Initial machinery assumptions.

Four motorbikes (USD)	$5,000
Small baler (USD)	$5,000
Small wash (USD)	$5,000

As value is added to the recycled plastic, its price per kilo is increased. These are based on values given by our Ghanian Entrepreneur partners, though they are heavy estimates and do not factor the point of sale price negotiations (Table 2.7).

Table 2.7: Plastic price per kilogram.

Value added	Plastic price per kilogram (GHc)
Sorted	0.65
Cleaned + sorted	0.8
Baled + cleaned + sorted	0.9

To estimate the price of bins and bags, a quote from jiji.com was taken (Table 2.8).

Table 2.8: Bin system prices.

	Cost (USD)	Cost (GHc)
Bin	$60	390
Bags (100 packs)	$30	195

Table 2.9 contains price quotes for the machinery needed to run the Makerspace operation.

Table 2.9: Makerspace machinery quotes.

	Cost (USD)	Cost (GHc)
Hydraulic press (100-ton force)	$17,000	1,10,500
Sand poly extruder	$28,000	1,82,000
Mold (at least 2 sets)	$16,000	1,04,000
Cooling system	$4,000	26,000
Plastic crusher	$12,000	78,000
Washing system	$12,000	78,000
Training and consultation	$20,000	1,30,000
Bailer	$15,000	97,500

These ROI calculations serve as a good preliminary estimation of rate of return, final profit, time until reimbursing the initial investment, and the ratio of profit to initial investment. However, as previously noted, the virtual nature of this project required a considerable number of assumptions to be made. Lacking information about the size and weight of full bags, the volume of transportation trucks, and others resulted in imperfect equations.

Looking forward, we suggest incorporating the volume reduction from the baler as its own separate value. All value-added innovations are factored in by increasing the plastic price per kilogram value. For example, gaining the small baler increased plastic price per kilogram from 0.8 to 9 c/kg. It is estimated that this is reflective of the value gained by adding a baler; however, it is an indirect means of calculation.

Three financial methods were proposed to the chiefs. The first was low investment, fast return. This method involved only a $660 initial investment in bins for each village, assuming Dwenase already had bins, and would begin seeing profit after only 0.24 years. After 10 years, this method would earn $81,920 at a rate of $161.35/week with a profit to initial investment ratio of 4137.42%. The second method was characterized as the high initial investment model. This began with the purchase of bins in each village but also featured the purchase of motorbikes, a bailing system, and a cleaning system in Akyem Dwenase. The initial investment of $33,980 would be paid back in full after 3.1 years, earning $75,615.54 at a rate of $210.76/week with a profit to initial investment ratio of 222.53%. The final method is a continuous reinvestment model nicknamed "Hybrid." This system starts by following the low investment model for the first year until enough money has been earned to pay back the initial investment and purchase four motorbikes for $5,000. This process is continued with a $5,000 cleaning system and a $5,000 small baler, immediately reinvesting profits back into the business until the maximum rate of return is achieved. The Hybrid system returns its initial investment back after 0.24 years, earning a total of $80,750.81 after 10 years at a final rate of $210.76/week with a profit to initial investment ratio of 4,078.32% (Table 2.10).

Table 2.10: Outputs of each method.

Method	ROI	Initial investment, USD	Profit 10 years, USD	Weekly rate of return 10 years, USD	Ratio of profit to initial investment
Low	0.24	$1,980.00	$81,920.89	161.3478591	4,137.42%
High	3.10	$33,980.00	$75,615.54	210.760654	222.53%
Hybrid	0.24	$1,980.00	$80,750.81	210.760654	4,078.32%
Method	ROI	Initial investment, Ghc	Profit 10 years, Ghc	Weekly rate of return after 10 years, Ghc	Ratio of profit to initial investment
Low	0.24	12,870.00	532,485.76	1,048.761084	4,137.42%
High	3.10	220,870.00	491,501.01	1,369.944251	222.53%
Hybrid	0.24	12,870.00	524,880.27	1,369.944251	4,078.32%

After presenting these three methods to the village chiefs, they noted a preference in the Hybrid model. As such this is the system that we will implement. In Figure 2.11, the Hybrid investment model is inferior to the low investment model. However, what the data shown here omits is the opportunity to scale and collect more plastic. If larger amounts of plastic were to be collected, the added value per kg of recycled plastic that

ROI Calcilations

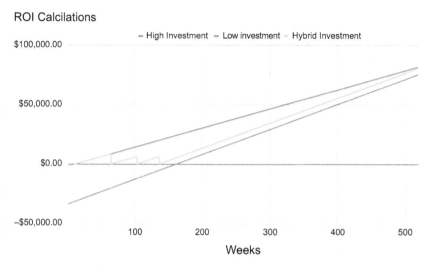

Figure 2.11: ROI comparisons: to better visualize the value of each method we have displayed their returns over a 10-year period.

the cleaning system and binning machine provide would have a greater effect. This is what the chiefs are betting on. They believe that over the course of the recycling initiative, the program will expand and recover more plastic. It must be noted that scaling the business may not always be the best course of action. If a village is unable to expand their collection, investing in a greater value per kg of recycled plastic may not be worth it. Lastly, the effect of our required assumptions cannot be overstated. The values presented were calculated off the information available to us, but we believe more accurate values could be found.

References

[1] Ritchie, H. & Roser, M. (2018, September 1). Plastic pollution. Our World in Data. Retrieved February 21, 2022, from https://ourworldindata.org/plastic-pollution

[2] Addaney, M. & Anarfiwaah Oppong, R. (2015). Critical Issues of Municipal Solid Waste Management in Ghana. *Jenrm, 2*(1), 30–36.

[3] Clapp, J. (2002). *The Distancing of Waste: Overconsumption in a Global Economy. Confronting Consumption.* Cambridge, MA: MIT Press, 155–176. https://mitpress.mit.edu/books/confronting-consumption

[4] Leal, M., Jorge, S. P., Charbuillet, C., & Perry, N. (2020). Design for and from Recycling a Circular Ecodesign Approach to Improve the Circular Economy. *Sustainability, 12*(23), 9861. https://doi.org/10.3390/su12239861

[5] Overview. World Bank. (n.d.) .Retrieved February 21, 2022, from https://www.worldbank.org/en/country/mic/overview#1

[6]   Chen, D. M.-C., et al. 2020. The World's Growing Municipal Solid Waste: Trends and Impacts. *Environmental Research Letters, 15*(7), 074021. https://doi.org/10.1088/1748-9326/ab8659

[7]   Agyei-Mensah, S. & Oteng-Ababio, M. (2012). Perceptions of Health and Environmental Impacts of E-waste Management in Ghana. *International Journal of Environmental Health Research, 22*(6), 500–517. 10.1080/09603123.2012.667795

[8]   Williams, M., Gower, R., Green, J., Whitebread, E., Lenkiewicz, Z., & Schröder, P. (2019). No time to waste: Tackling the plastic pollution crisis before it's too late.

[9]   Costas, A. (2021). Velis and Ed Cook Mismanagement of Plastic Waste through Open Burning with Emphasis on the Global South: A Systematic Review of Risks to Occupational and Public Health. *Environmental Science & Technology, 55*(11), 7186–7207. 10.1021/acs.est.0c08536

[10] Lissah, S. Y., et al. Managing Urban Solid Waste in Ghana: Perspective and Experiences of Municipal Waste Company Managers and Supervisors in an Urban Municipality. 2020. https://doi.org/10.21203/rs.3.rs-65452/v1.

[11] Ofori-Boateng, C., Teong Lee, K., & Mensah, M. The Prospects of Electricity Generation from Municipal Solid Waste (MSW) in Ghana: A Better Waste Management Option. Fuel Processing Technology. Elsevier, January 21, 2013. https://www.sciencedirect.com/science/article/pii/S037838201200447X?via%3Dihub#bb0130.

[12] Landfills: Investigating Its Operational Practices in Ghana. Accessed March 2, 2022. https://www.researchgate.net/publication/338712210_Landfills_Investigating_Its_Operational_Practices_in_Ghana.

[13] Africa Waste Management Outlook Summary, UN Enviroment,https://wedocs.unep.org/bitstream/handle/20.500.11822/25515/Africa_WMO_Summary.pdf?sequence=1&isAllowed=y

[14] Visual Feature: Beat Plastic Pollution. UNEP. Accessed March 2, 2022. https://www.unep.org/interactive/beat-plastic-pollution/.

[15] World Bank Group. Solid Waste Management. World Bank. World Bank Group, February 11, 2022. https://www.worldbank.org/en/topic/urbandevelopment/brief/solid-waste-management.

[16] Letcher, T. M. (2020). Introduction to Plastic Waste and Recycling. *Plastic Waste and Recycling*, 3–12. https://doi.org/10.1016/b978-0-12-817880-5.00001-3

[17] Recycling Guide: Plastic, Glass, Metal, Paper, and More. Vivbizclub, https://vivbizclub.com/recycling-guide-plastic-glass-metal-paper-and-more/.

[18] Teta, L., et al. *Poor Plastic Waste Management in Accra*. Ghana.

[19] Accra Metropolitan Assembly composite budget for the 2015 fiscal year. Republic of Ghana.

[20] Kumasi Metropolitan Assembly composite budget for the 2016 fiscal year, Republic of Ghana.

[21] Miezah, K., Obiri-Danso, K., Kádár, Z., Fei-Baffoe, B., & Mensah, M. Y. (2015). Municipal Solid Waste Characterization and Quantification as a Measure Towards Effective Waste Management in Ghana. *Waste Management, 46*, 15–27.

[22] Stoler, J., Weeks, J. R., & Fink, G. (2012). Sachet Drinking Water in Ghana's Accra-Tema Metropolitan Area: Past, Present, and Future. Journal of Water, Sanitation and Hygiene for Development: A Journal of the International Water Association, U.S. National Library of Medicine, https://www.ncbi.nlm.nih.gov/pmc/articles/PMC3842094/#:~:text=Sachet%20water%20typically%20consists%20of,and%20other%20West%20African%20nations

[23] Coletto, D. & Bisschop, L. (2017). Waste Pickers in the Informal Economy of the Global South: Included or Excluded?. *International Journal of Sociology and Social Policy, 37*(5/6), 280–294. https://doi.org/10.1108/IJSSP-01-2016-0006

[24] Meyerhoff, R. SAP Brandvoice: Ghana's Ambitious Plan to Minimize Plastic Waste. Forbes. Forbes Magazine, October 22, 2020. https://www.forbes.com/sites/sap/2020/10/22/ghanas-ambitious-plan-to-minimize-plastic-waste/?sh=88795bea3274.

[25] List of Recycling Companies in Ghana. (2022/2023). June 11, 2021. https://myhealthbasics.site/list-of-recycling-companies-in-ghana/.

[26] Magoum, I. Ghana: A New Solid Waste Recycling Initiative in Accra. Afrik 21, 9 Sept. 2021,
      https://www.afrik21.africa/en/ghana-a-new-solid-waste-recycling-initiative-in-

[27] Boateng, N. Nelplast Ghana Ltd, http://www.nelplastgh.com/.

[28] Eglash, R. (2016). An Introduction to Generative Justice. *Teknokultura, 13*, 369–404. 10.5209/
      rev_TEKN.2016.v13.n2.52847

[29] Eglash, R. & Garvey, S. (2014). Basins of Attraction for Generative Justice10.1007/978-94-017-8691-1_5.

[30] Barton, P. A. (2001). *Susu & Susunomics: The Theory and Practice of Pan-African Economic, Racial and
      Cultural Self-Preservation*. New York: Authors Choice Press.

Adam Dincher, Karris Krueger, Casey Snow, Kelvin Atograph,
Joseph Atedoghu, Fiifi Addae, Ayomide Akerejola, Nelson Boateng,
Kwasi Anane Asare, Osabarima Owusu Baafi Aboagye III,
Hermine Vedogbeton, Robert Krueger

Chapter 3
Utilizing bricks made from recycled plastics to construct a gravity drip irrigation system in Akyem Dwenase, Ghana

3.1 Introduction

Global climate change is a phenomenon that affects the world's regions in different ways. According to the United Nations, West Africa is set to become a climate-change hotspot [1]. In West Africa, these effects come in the form of higher temperatures, increased climate variability, desertification, and drought [2]. Historical data from 1960 to 2000 shows a progressive rise in temperature in the region. In these 40 years, the average temperature increased by 1 °C [3]. In the Sahel region of sub-Saharan Africa, where 80 of 100 million people live on < $2/day, climate change is bringing sea-level rise, reducing already stretched water resources, and decreasing agricultural yields. These impacts on resources have contributed to loss of livelihoods, civil unrest, and violence across the Sahel, which stretches from Senegal to the west and Eritrea to the east.

The country of focus here is Ghana in West Africa. Ghana faces similar climate change impacts as Sahel, which is to the north, such as increased evaporation rates, diminished water resources, and, in Northern Ghana, desertification. Indeed, each of Ghana's six agroecological regions, from the desert environments to the north and the tropical ecosystem to the south, has been affected by climate change, resulting in an

Adam Dincher, Karris Krueger, WPI student, Architectural Engineering
Casey Snow, WPI student, Computer Science
Kelvin Atograph, E-waste processor and computer repairer, Agbogbloshie, Ghana
Joseph Atedoghu, Fiifi Addae, Ayomide Akerejola, ACUC student, Engineering
Nelson Boateng, President, Nelplast, LtD
Kwasi Anane Asare, Agricultural Chief, Akyem Dwenase, Akyem Abuakwa Traditional Area
Osabarima Owusu Baafi Aboagye III, Chief, Akyem Dwenase, Akyem Abuakwa Traditional Area
Hermine Vedogbeton, Social Science and Policy Studies Department
Robert Krueger, Director, Institute of Science and Technology for Development

https://doi.org/10.1515/9783110786231-004

average decrease in annual rainfall. The main rainy season in Ghana, between May and August, accounts for 75% of the total annual rainfall for the country. The major dry season falls between mid-November to April, with a short dry spell in July and August. The average rainfall for the entire country for this season was ~ 930 mm (1960–1990) and has dropped to ~ 787 mm on average (1990–2020) (https://climateknowledgeportal. worldbank.org/country/ghana/climate-data-historical, accessed 7/1/22).

Diminished rainfall resulting from a changing climate has limited agricultural output, in particular, and the region's flora, in general. The dry, semiarid climate in the Northern region leaves many farms at risk as the climate crisis evolves, but challenges exist in the Southern, more humid regions, too. With rainfed agriculture being a dominant economic and subsistence activity during the rainy seasons, it has become highly vulnerable to global climate change and climate variability [3]. These impacts provide specific challenges to communities in Ghana, where subsistence and market-oriented agriculture account for 60% of the total economic activity and any insufficiencies can put a village's food security at risk [4]. Climate variability and lower annual rainfalls have put West Africa at risk for economic crisis as well as an increased risk for food insecurity. The timing of the rains has real implications for farmers in Ghana – and everywhere without a source of anthropogenic irrigation. In their study of Acute Food Insecurity, Shukla et al. [11] point out that rain's onset dates are critical when it comes to labor availability, purchasing seed and fertilizers, and choosing planting times.

3.1.1 Design challenge

There are a number of ways that scientists and engineers are intervening in this issue from using machine learning to massive infrastructure projects, such as dams and public work projects. In contrast, this project engages the water shortage issue and its accompanying impacts at the village scale. This project, which is ongoing, seeks to ameliorate the impacts of climate variability by introducing gravity-fed irrigation to the region. North of the coastal area, much of Ghana has undulating terrain that makes it amenable to gravity-fed systems. These systems are cheaper and offer more user-friendly maintenance than systems with gasoline or electric pumps. Though not high-tech, these systems can be part of a smart village scheme that achieves the goals of this volume: to use local forms of knowledge and establish collaborative projects at an appropriate scale so they create opportunities for generative justice and, by extension, self-sufficiency.

To develop our proof of concept, we partnered with a community located in the Eastern region of Ghana, the town of Akyem Dwenase. We worked with Chief Osabarima Owusu Baafi Aboagye III, and his Agricultural Chief, Kwasi-Anane Asare, who manages the chief's agriculture operations, as well as other residents. Our site is a 16-acre field (Figures 3.1 and 3.2) that Osabarima established to produce cocoa, a cash crop for export. Currently, this field relies on rainfall as its sole source of water. Here is where climate variability is most problematic: if the rainy season comes early or late, the cocoa yield can diminish because a plant's resistance to drought is determined by when it is introduced. If the rains come too early, the plant is at greater risk of being immature when the rains leave, thus increasing the chance that it will die before the rains return. If the rains come too late, the juvenile plants will wilt and die before the rains come.

To ensure crop's success, the fields must now have access to a reliable water source to last throughout the dry season. Together with the agricultural chief and supplier bricks made from recycled plastic, Nelson Boateng, and four engineering students from Academic City University College (ACUC) in Accra – Kelvin Atograph, Joseph Atedoghu, Fiifi Addae, and Ayomide Akerejola – we collaborated from January to March 2022 to design a gravity-fed, drip-irrigation system made of locally available materials, including a structure to be made from recycled plastic. The remainder of this chapter describes this process and links the project to the broader goals of this collection.

Figure 3.1: Akyem Dwenase's cocoa field landscape.

3.2 Essential background

Ghanaians have been producing cocoa crops since the nineteenth century. Since then, cocoa has become Ghana's largest agricultural export commodity [5] and while the country is the current second largest producer in the world, for much of the twentieth century, it was the world's largest producer of cocoa. Despite contributing so much to the chocolate industry, Ghanaians only recently have sought to add value to this mainstay through the boutique chocolate industry.

As with all agriculture in West Africa, cocoa farmers are affected by global climate change [6]. The factors that are most common are low soil moisture and excessive exposure to sunlight. Exacerbating this is the many rural Ghanaians who live on a median income of $60 per year for cocoa farming [7]. Often, Ghana's cocoa farmers have limited access to finances that would allow for the purchase of specific supplies and planting materials to address deficiencies. While researchers estimate that up to 40% of the crop is lost to pests and disease [8], < 2% of the total cultivated area in Ghana is irrigated; it is certain that more encouragement to progress irrigation development is needed [9]. With more than 70% of Africa's most vulnerable populations relying on agricultural production for their livelihoods, an irrigation system design that promotes self-sufficiency and sustainability is clearly needed.

On this score, we sought to incorporate locally produced building materials. Our partner, Nelson Boateng, President of NelPlast Eco Ghana, Ltd, has, for almost a decade, utilized recycled plastic waste to produce materials for construction. His design allows for these bricks to stay cool in Ghana's warm climate while also being fire resistant. The LEGO-like design and implementation of rebar-sized holes resolve the need for any special adhesive or mortar component to ensure structural integrity. Moreover, NelPlast has been creating structural bricks for roads, parking lots, and, most recently, residential homes all over Ghana. Using a 70:30 ratio of plastic to sand creates an extremely durable brick that can withstand the arid climate and is cheaper than importing raw materials.

Until this project, Boateng's blocks were not used for an agricultural purpose. He shared that he never used the bricks for other applications, outside of housing structures and pavement substitutes. We inquired if he believed the bricks could be used to support a water reservoir. Though the plastic bricks alone are waterproof, they must be sealed to become impermeable and hold water in, as opposed to keep it out.

In the subsequent sections of this chapter, we examine the different components of irrigation systems, their design, and need for co-design, especially as we reclaim the concept of "smart villages."

3.2.1 A closer look at irrigation systems

Rain-fed systems are a primary source of crop production, which in turn limits most of the crop to a 3- to 6-month rainy season. With a water retention method, farmers could cultivate year-round, mature crops faster, and grow higher value crops that require a more reliable water source. Implementing reliable water sources such as an irrigation system can mitigate the impacts of climate stress associated with expected drought and extreme heat [12].

Irrigation systems are established all over the world and utilize unique characteristics based on their needs and geographical circumstances. The methods native to Africa's western sub-Saharan climate are chosen based on numerous factors. These include elements like the amount of available rainfall in a season, crop type, soil type, environmental conditions, and topography. The most common systems are based around manual irrigation, the traditional distribution of water through manual labor, and watering cans to supplement any lack of rainfall. A system like surface irrigation uses the existing structure of the land and gravity to distribute water with no mechanical pump involved. When looking to invest in long-term crop growth, some farmers choose to implement sprinkler irrigation with pumps that allow water to be quickly distributed overhead from a high-pressure sprinkler.

Drip irrigation is a localized form of irrigation in which water is dripped at or near the root of the crops. This method is satisfactory for farms located on mountainous or sloped topography as gravity can be used to induce water pressure. Another desirable factor for drip irrigation is the consistent and direct feed of water going to plants allows them to grow more efficiently. This form of watering is preferable for arid climates, where much of the surrounding vegetation is competing for root space and moisture in the soil [10]. In terms of water distribution, there were two main concerns: determining the total amount of water needed to feed the cocoa trees and designing a distribution system to deliver the right quantity of water to each tree.

There are maintenance concerns with drip irrigation. The lines can clog, fail under environmental conditions, and are inflexible in terms of supporting multiple crops with varying water needs. The lines exposed to the soil can cause potential blockage in the pipes, stopping the water's flow as well. Another issue with drip irrigation is that it does not adhere to each specific plant's watering needs. These concerns aside, with proper training for initial construction and repair and maintenance, a drip irrigation system can serve all the project's goals.

3.2.1.1 Irrigation system materials

When considering what materials to construct with, we identified that materials coming in direct contact with the funneled water need to be water-resistant, sustainable, and readily attainable in Ghana. Polyethylene comes in a variety of forms and densities and is commonly used in many agricultural industry applications. It has a high resistance to corrosion and exceptional toughness that can stand up to harsh conditions. Additionally, materials like mesh screen filters, supporting garden stakes, rain chains, and grommets are all metals that could be subject to corrosion. In our case, we used bronze and stainless steel materials that are heavily resistant to corrosion and do not rust with moisture. Metals like aluminum that are found in the repurposed garden stakes can be protected by a polyethylene layer.

3.3 The process

At its core, a drip irrigation system has three main components: a catchment system to collect the rainwater, a tank or reservoir to store the collected water, and piping connections to deliver the water. This section discusses the technical and social process of design we employed.

3.3.1 Rainwater collection

With our partners, we decided a catchment system that funneled water into a reservoir was the best choice. To ensure we had enough water, we needed to determine the surface area required to collect the right amount of water and the size of the reservoir.

To establish the necessary surface area, we first wanted to find how much water would be necessary to survive a dry season. The agricultural chief said that each cocoa plant needs 25–50 mm of water per week. With the typical dry season lasting around 16 weeks, we planned for 813 mm of water per plant. This led us to opting for a 3,800-L tank. To collect the amount of water needed to fill this tank, the surface area of the catchment must be ~ 2.5 m.

To design the most appropriate catchment system for the cocoa farm in Dwenase, the team decided on a conical-shaped surface prototype for rain to land on, which funnels into the reservoir (Figure 3.2). Local suppliers could support this design as it is scalable and can be developed into a small business opportunity.

Figure 3.2: An example of a preexisting, small-scale rainwater collection system.

To design the system in prototype form, we worked at a 1:10 scale. Our catchment design employs an inverted umbrella-like shape to feed rainwater into the water tank. Using a square cutout of polyethylene drop cloth material (2 mil), we shaped the rain collection surface. The catchment's polyethylene sheet is supported by a structure designed from garden stakes (Figure 3.3a). These garden stakes, made from polyethylene-coated hollow aluminum tubing, created a stable secure structure between the garden stakes and polyethylene. Then, we installed grommets at the corners of polyethylene and attached it to the stakes. To create a support ring underneath the outer surface, we used smaller plastic stakes (Figure 3.3b). It was easy to create more support as needed by connecting straight sections of any material between the existing supports. To produce a center feed, we installed a grommet that allows the catchment layer to hook onto the connection that leads into the tank.

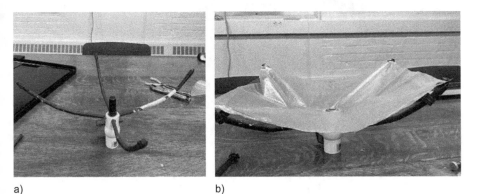

a)   b)

Figure 3.3: (a) Catchment prototype structure (left), and (b) tarp layout (right).

Forming the connection between the surface and the reservoir is a central coupling connection with a primary filter on the top (Figure 3.4), and a three-layer secondary filter on the bottom. These filters prevent silt and other debris from entering the irrigation piping and causing clogging.

The last step was adding a "rain chain." A rain chain guides the rainwater from the mesh filter down to the water surface (Figure 3.5). This prevents the first flush of water from mixing sediment that may have washed in from the catchment sauce.

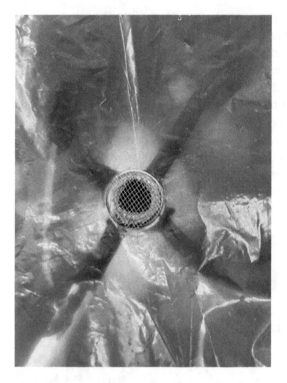

Figure 3.4: Primary filter.

3.3.2 The reservoir

Remember, we determined the reservoir size needs to hold 3,800 L of rainwater. The reservoir will be constructed using NelPlast's bricks. Given the size and weight of these bricks, it was not feasible to obtain the bricks in the United States. Fortunately, partners from the ACUC in Ghana could obtain the bricks. Our ACUC partners tested the bricks using a sealed section filled with water on the seam of two connected bricks. When water was applied, the seam let water through, which we all suspected. The ACUC students then employed a piece of polyethylene to cover the sidewall and seam of the bricks. After testing the bricks (see Figure 3.6) with the polyethylene sheet, our partners noted that no water had escaped through or damaged the material.

The next step was taking the information collected by the Ghanian team to continue with the prototype. We employed NelPlast's design for home construction into the reservoir design. We determined this direction after examining the LEGO Brick and Pilar Brick's uses in building interior and exterior walls of houses. The team agreed to stagger the bricks so that each row was offset by half a brick in the horizontal direction. Through research we created initial designs (Figure 3.7) and drew them on *SolidWorks*. The LEGO Bricks were estimated through digital photo measuring 6 in wide × 6 in tall × 12 in long.

Figure 3.5: Rain chain.

Figure 3.6: NelPlast's "LEGO" brick mold.

Figure 3.7: Initial SolidWorks' design.

The corner brick dimensions were 6 in wide × 8 in tall × 4 in long. From this CAD model, we were able to 3D print them in our school's makerspace.

From this first design iteration, we printed brick models. We made the extrusions the same size as the protrusions. With that in mind, we developed new LEGO and column bricks. The new ones had the same dimensions; however, the extrusions were 0.1 in wider, and the protrusions were 2.54 mm narrower. We then printed the new blocks and evaluated them. The fit was still not as seamless as we had hoped (Figure 3.8).

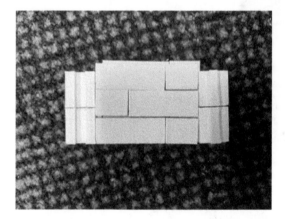

Figure 3.8: Second brick iteration prototype.

Working based on assumptions of the bricks failed to produce a good prototype. The team in Ghana acquired and measured the bricks.

After replicating the exact dimensions in SolidWorks, we printed the redesigned bricks again (Figure 3.9). These fit slightly better, but still needed their margin of error increased due to the variability of the 3D printer. We increased the extrusions to be 5 mm wider, and the protrusions to be 10 mm narrower, making for a total of 5 mm between the bricks. Once these were printed, the prototype bricks fit together as actual bricks (Figure 3.10).

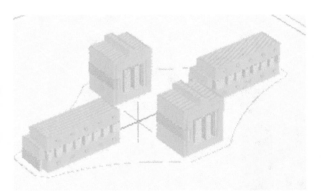

Figure 3.9: Bricks on 3D printer software.

3.3.3 Water distribution system: the tubes and pipes

The prototype for testing the tubing did require the use of 3D printed bricks. Because this part of the design process focused on testing the pipe and water delivery mechanism, we used a recycled milk jug (Figure 3.11). We used a drill bit and X-Acto knife to carve our feed hole into the milk jug to fit a ball valve inside. To secure the ball valve, we hot glued the seams and then added a layer of hardening epoxy to prevent any small leaks from getting through the hot glue barrier.

Figure 3.10: Final 3D printed bricks assembly.

Figure 3.11: Initial tubing prototype.

From the research described above, we knew different tubing sizes were necessary for our design. The larger tube directly attached to the tank we used was half-inch PVC piping. In our first design, we laid the PVC pipe perpendicular to the edge of the tank and run down the field (Figure 3.12a). After some testing, we ran the pipe parallel to the front edge of the tank (Figure 3.12b). The parallel design allows for more variability, as well as the capability to water more plants.

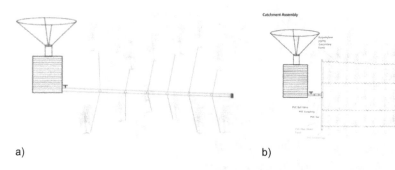

a)   b)

Figure 3.12: (a) Initial perpendicular pipe (left), and (b) parallel pipe (right).

For the smaller pipes, in our first prototype iteration, we used ⅜ in vinyl piping and drilled holes to test the delivery flow. After initial tests, we found that the holes were too big. Then, partners in Ghana suggested using polyethylene piping instead of vinyl. The next iterations consisted of using polyethylene piping with smaller holes. This time the holes were poked through with a push pin and worked significantly better – more like how they would in the field.

3.3.4 Replicating the farm

Because the project was remote (i.e., we were not physically in Ghana), the process of replicating field conditions was difficult. We had drone footage and aerial photography that we had commissioned. So, we started with what we knew. The farm is 16 acres, and the trees are laid out in a 2.5 m × 2.5 m grid. We also knew that Nana Kwasi was cultivating four different crops: cocoa, cassava plants, plantains, and palms. What was unclear from the drone footage was the precise topography. Thus to account for unknown variability of the field, we designed our prototype to be dynamic.

To properly replicate the farm, the soil was necessary to see if the pipes clogged when they were in the dirt. To achieve this, we placed dirt into a large plastic container and set up the prototype in it. Although this plastic bin is nowhere near the size of the farm itself, we needed to scale it down for practicality and feasibility purposes. The pipes did not clog, and we were easily able to direct them to certain locations in the dirt. We purposely used a flexible material for the smaller pipes to account for the fact that the cocoa plants may not be in an exact grid. We also sloped the dirt in a way that angled the pipes downhill. This was to replicate the field's slope and to encourage the gravity-fed aspect of the system.

The final prototype was assembled and ready to be tested (Figure 3.13).

3.3.5 System maintenance

In case the system failed, we examined several different maintenance options. One of those was gate valves along with the PVC (Figure 3.14). If one section was not working properly, a gate valve could close off that section during repairs. These gate valves also could provide a different way for the water to pressurize if one section was not getting enough pressure, or a section was flooding.

Additionally, in a meeting with our ACUC partners in Ghana, they suggested if the bricks were not successful, we should consider using old refrigerators. Many refrigerators can be found in the e-waste dumps, and while they do not work electronically, they still have their sturdy frame and function that could be repurposed. In terms of sealing water in, a fridge is designed to keep air, and subsequently moisture, within its body. During the rainy season, the fridge would simply be placed on its back and filled with water. When the dry season comes, the refrigerator door could be closed and it would seal out a large amount of outside heat and prevent evaporation. These "broken" refrigerators are easily found at dumpsites, highly affordable, and in theory, serve us well in their repurposed use (Tables 3.1).

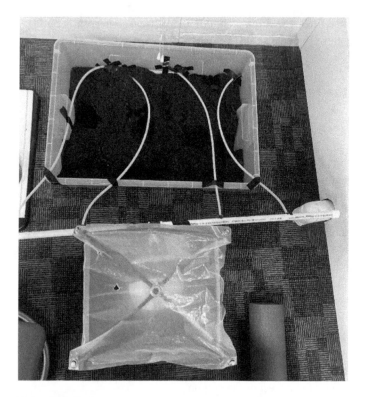

Figure 3.13: Final prototype.

Figure 3.14: Gate valve design.

Table 3.1: Prototype materials.

	Average of Trials (empty line)		
	Water Amount (mL)	Time (s)	Flow Rate (mL/s)
Cup 1	6.0	31.85	0.19
Cup 2	6.5	31.85	0.20
Cup 3	4.4	31.85	0.14
Cup 4	20.0	31.85	0.63
Cup 5	61.0	31.85	1.92
Cup 6	4.0	31.85	0.13

3.4 Testing and analysis of the data

Our prototype tested two different situations: (1) letting the pipes continue to drip after the spigot was shut off versus moving the cups once the spigot was shut off to only measure what amount of water was distributed during the time the spigot was open; and (2) running the test when the pipes were dry versus running the test immediately after another when there was still water standing in the lines. The most realistic conditions to replicate the actual field would be to run the water on dry pipes and let it continue to drip after being shut off. In addition to these situations, we also tested several different variables: (1) small versus large holes in the tubes; (2) uphill versus downhill locations; (3) one hole versus two holes on the line – to simulate few or many plants on a run; and (4) if the line was closer to the tank or farther down the main PVC pipe from the tank. Tables 3.2 and 3.3 display and summarize these results. There was a large difference in data as a cup placed alone on a line with a larger than normal hole farthest from the tank had an average of 44.75 mL and a high of 61 mL while a cup farthest from the tank, downhill, with average sized holes, and two cups on the line had an average of 6 mL and a low of 4.4 mL.

Table 3.2: A matchup of different variables tested (data collected from trial 1 only in this case) to see what combination created the most ideal results.

	Average of Trials (moved cups)		
	Water Amount (mL)	Time (s)	Flow Rate (mL/s)
Cup 1	7.10	33.65	0.21
Cup 2	7.75	33.65	0.23
Cup 3	3.80	33.65	0.11
Cup 4	7.40	33.65	0.22
Cup 5	39.00	33.65	1.16
Cup 6	5.10	33.65	0.15

We ran four roughly 30 s trials and measured the amount of water collected in each cup after using a syringe and a graduated cylinder. At first, we assumed cup 5, which gave us a measurement of 61 mL in our ideal conditions, was the best set of variables to choose as the end product. However, when we calculated the flow rate of 1.92 mL/s (61 mL/31.85 s), we realized that for a goal of watering 1 in (16.67 mL), we only needed the system flowing for approximately 9 s. This would be too short a time span to build pressure in the pipes enough to water other plants in the line or to control the amount in a reliable way.

Table 3.3: Average for each cup over all four trials.

	Average of Trials (drip after)		
	Water Amount (mL)	Time (s)	Flow Rate (mL/s)
Cup 1	6.10	31.845	0.19
Cup 2	11.25	31.845	0.35
Cup 3	8.20	31.845	0.26
Cup 4	19.00	31.845	0.60
Cup 5	50.50	31.845	1.59
Cup 6	7.75	31.845	0.24

Further data analysis showed that the best option to choose from the prototype test for our actual irrigation system would be first to have many plants on the line, not few, which works in our favor to allow us to use less materials in the end, thus lower costs; and second to make smaller holes in the tubing lines to create a slower drip and slower flow rate. Having a slower flow rate means the water will drip at a more controlled rate, so the soil can absorb more moisture more quickly without being flooded and risk water flowing away from the plant and taking valuable soil and nutrients away from it as well. The cups that best show these conditions and setup are cups 1–4 (Figure 3.15). The average of these four test locations from trial 1 is 9.23 mL, as given in Table 3.4 (two holes × two holes).

Figure 3.15: Labeled cups.

Table 3.4: Raw data from each of our four trials for each cup with a nominal description to the left of the data.

				Trial 1		Trial 2		Trial 3		Trial 4	
				Water Amount (mL)	Time (s)	Water Amount (mL)	Time (s)	Water Amount (mL)	Time (s)	Water Amount (mL)	Time (s)
Top	Far Side	Pipe 1	Cup 1	6	31.85	6.2	31.84	7.8	35.87	6.4	31.43
Bottom	Far Side	Pipe 1	Cup 2	6.5	31.85	16	31.84	8	35.87	7.5	31.43
Top	Close	Pipe 2	Cup 3	4.4	31.85	12	31.84	3.6	35.87	4	31.43
Bottom	Close	Pipe 2	Cup 4	20	31.85	18	31.84	8.7	35.87	6.1	31.43
Larger Hole	Far Side	Pipe 3	Cup 5	61	31.85	40	31.84	42	35.87	36	31.43
	Close	Pipe 4	Cup 6	4	31.85	11.5	31.84	5	35.87	5.2	31.43

Table 3.5 depicts the differences in data between letting the system drip into the cups after being shut off and moving the cups away from the holes to end cap the amount of water measured during the time span and no more.

Table 3.5: Prototype trial data.

	Average of All Trials		
	Water Amount (mL)	Time (s)	Flow Rate (mL/s)
Cup 1	6.60	32.75	0.20
Cup 2	9.50	32.75	0.29
Cup 3	6.00	32.75	0.18
Cup 4	13.20	32.75	0.40
Cup 5	44.75	32.75	1.37
Cup 6	6.43	32.75	0.20

	Other Variables Tested (mL/31.85sec)							
	One Hole	Two Holes	Small Hole	Large Hole	Uphill	Downhill	Close to tank	Far from tank
One Hole	32.5		4.00	61			4	61
Two Holes		9.23			5.2	13.25	12.2	6.25
Small Hole	4			8.18	5.2	13.25	12.2	6.25
Large Hole	61				61			61
Uphill		5.2	5.2			5.2	4.4	6
Downhill		13.25	13.25				13.25	6.5
Close to tank	4	12.2	12.2		4.4	20	9.47	
Far from tank	61	6.25	6.25	61	6	6.5		24.5

Table 3.6 depicts the difference in data between running the system and measuring the amount of water in the cups on an empty, dry line and measuring the amount of water after running the system with tubes already filled with water.

Table 3.6: Prototype trial data.

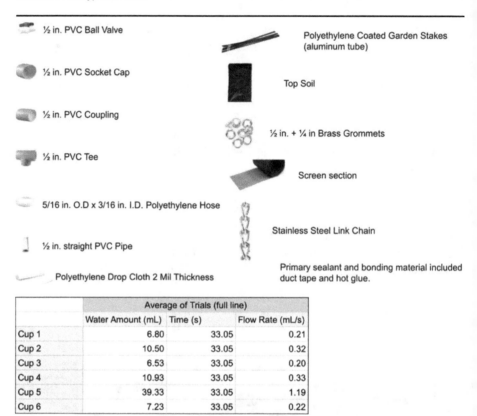

½ in. PVC Ball Valve

½ in. PVC Socket Cap

½ in. PVC Coupling

½ in. PVC Tee

5/16 in. O.D x 3/16 in. I.D. Polyethylene Hose

½ in. straight PVC Pipe

Polyethylene Drop Cloth 2 Mil Thickness

Polyethylene Coated Garden Stakes (aluminum tube)

Top Soil

½ in. + ¼ in Brass Grommets

Screen section

Stainless Steel Link Chain

Primary sealant and bonding material included duct tape and hot glue.

	Average of Trials (full line)		
	Water Amount (mL)	Time (s)	Flow Rate (mL/s)
Cup 1	6.80	33.05	0.21
Cup 2	10.50	33.05	0.32
Cup 3	6.53	33.05	0.20
Cup 4	10.93	33.05	0.33
Cup 5	39.33	33.05	1.19
Cup 6	7.23	33.05	0.22

3.5 Reflection on this project and the aims of the book

3.5.1 Smart villages

The village of Akyem Dwense can now be considered smart, not because it uses high-tech items for the betterment of the village, but because it uses an innovative solution. This creative solution uses gravity as its driving force. It improves the village, while also building on what is already somewhat successful. The cocoa farming methods are now working but could be improved, and we hope that our system will greatly enhance it.

Furthermore, the drip irrigation system will benefit the village of Akyem Dwense in a variety of ways. As previously stated, cocoa is Ghana's number one export crop, so it is in high demand. That means Chief Osabarima can sell the cocoa product from his farm and use the profit to the benefit of the village. He plans to improve the schools and value of education with the money these crops bring in. Without this system, there is less of a chance the cocoa plants survive, which in turn results in less yield. Less crop yield continues the chain reaction, ultimately not leading to the desired success with the system.

Additionally, the drip irrigation system will allow the farmers to spend less time watering their plants. With this extra time, farmers can focus on maintaining the field in other ways or even expanding. Expanding would allow more cocoa to be grown, which will result in more money for the village. This extra produce can help sustain the demands of the crop.

Not only can the irrigation system help the crops, but Nelson's bricks can also contribute to promoting behavioral changes concerning recycling. People can see that his plastic bricks can be used to store water and help their fields. They get something in return for recycling. This positive reward system will hopefully encourage recycling.

Lastly, with this proof-of-concept design, if other villages do not have the same materials, they can substitute with other things more readily available to them.

3.5.2 Co-design and generative justice

Co-design also plays a vital role in the design and development of smart villages. Along with the legitimacy of the project, co-design supports the concept of generative justice in that it creates a chance to seize an opportunity for self-sufficiency and sustainability – if people in the village do not know how this system is designed, or how it works, then it cannot be fixed. Also, if they did not want this system, it will not be used.

Implementing a model of generative justice allows us to innovate around monetary needs and apply solutions that communities can benefit from most. It is essential to our design process to share with our partners in Ghana the necessary steps and decisions to create adequate feasibility for future implementation. Our inadequate local experience in planning, design, and construction projects were supported enormously by village leaders and ACUC scholars in Ghana. Indeed, this project would not have been possible without an interest in sharing ideas, learning how to communicate effectively, and realizing that everyone brought a legitimate perspective to the process.

Much of the high cost associated with irrigation development comes from the involvement of costly expatriate expertise at nearly every stage of the project cycle. A lack of co-design would also set back the project's feasibility study that could lead to exorbitant design changes during the construction of the system. By implementing a design process focused on generative justice, we avoid the procurement of nonstandard equipment that may require special maintenance and service arrangements. By designing with our partners in Ghana, we hope that they will not be tied to any external funds, methods, or resources not within the vicinity of Ghana.

Many journals and research publications remain hyper-focused on big tech conglomerates that strive to make money from the next-generation smart city and village development. These projects may result in interventions that offer short-term benefits, but as time goes on and repairs are needed, or new, similar projects initiated, communities are left in the same cycle of dependency. As scientists and engineers, we must be wary of implementing specific products based on technological enterprises looking to contribute to these efforts. Implementation of technology that is not being sought to incorporate local production or preservation is only for the benefit of business interests. This project was integrated within the local community. In every step of the way, we worked together to identify locally available materials and expertise that will support the construction, maintenance, and creation of new systems. Not only does this added part of the design process meet the needs of the community, but it also helps transform the cycle of dependence into one of self-determination, self-sufficiency, and sustainability.

References

[1]   Shepard, D. (2019). Global Warming: Severe Consequences for Africa: New Report Projects Greater Temperature Increases. *Africa Renewal, 32*(3), 34–34. https://www.un.org/africarenewal/magazine/december-2018-march-2019/global-warming-severe-consequences-africa.

[2]   Druyan, L. M. (2011). Studies of 21st-Century Precipitation Trends over West Africa. *International Journal of Climatology, 31*(10), 1415–1424.

[3]   Klutse, N. A., Owusu, K., Nkrumah, F., & Asa Anang, O. Projected Rainfall Changes and Their Implications for Rainfed Agriculture in Northern Ghana. *Weather, 76*(10), 340–347. https://doi.org/10.1002/wea.4015.

[4]   Frenken, K., ed. (2005). *Irrigation in Africa in figures: AQUASTAT survey, 2005*, Vol. 29. Food & Agriculture Org.

[5]   Kolavalli, S. & Vigneri, M. (2011). Cocoa in Ghana: Shaping the Success of an Economy. *Yes, Africa Can: Success Stories from a Dynamic Continent, 201*, 258643–1271798012256.

[6]   Schroth, G., et al. (2016). Vulnerability to Climate Change of Cocoa in West Africa: Patterns, Opportunities and Limits to Adaptation. *Science of the Total Environment, 556*, 231–241.

[7]   Hainmueller, J., Hiscox, M., & Tampe, M. (2011). *Sustainable Development for Cocoa Farmers in Ghana.* Cambridge (MA): MIT and Harvard University.

[8]   Tackie, E., Sheilo Baghr, P., Bisseleua, H. D., & Akwetey-Kodjoe, N. (2022, February 25). African Cocoa Initiative II. World Cocoa Foundation. https://www.worldcocoafoundation.org/initiative/african-cocoa-initiative-ii/.

[9]   Danish Ministry of Foreign Affairs, UNEP/UNDP. (February 2015). *National Climate Change Adaptation Strategy.* UNDP Climate Change Adaptation. https://www.adaptation-undp.org/mainstreaming-adaptation.

[10]   Büyükcangaz, H., Alhassan, M., & Nyenedio Harris, J. Modernized Irrigation Technologies in West Africa. *Turkish Journal of Agriculture – Food Science and Technology, 5*(12), 1524. https://doi.org/10.24925/turjaf.v5i12.1524-1527.1429.

[11] Shukla, S., Husak, G., Turner, W., Davenport, F., Funk, C., Harrison, L., & Krell, N. (2021). A Slow Rainy Season Onset is a Reliable Harbinger of Drought in Most Food Insecure Regions in Sub-Saharan Africa. *Plos one*, *16*(1), e0242883.

[12] Burney, J. A., Naylor, R. L., & Postel, S. L. (2013). The Case for Distributed Irrigation as a Development Priority in Sub-Saharan Africa. *Proceedings of the National Academy of Sciences*, *110*(31), 12513–12517.

Grace Fitzpatrick-Schmidt, Helen Le, Ethan Wilke, Nicholas Masse

Chapter 4
Women's health case study

4.1 Project description

Open-air burning of valuable metal from electronic components releases toxic contaminants such as dioxins and heavy metals, including lead, into the environment. These contaminants cause reproductive harm, cancer, and respiratory issues, specifically in women. These dangers are intensified for pregnant women, who transmit the toxins to their unborn children leading to birth defects, low birth weight, stillbirths, preterm birth, and many other long-term mental and cognitive problems [1].

There is a lack of awareness of the long-term health impacts posed by e-waste, and "when people fail to be alarmed about a risk or hazard, they do not take precautions" [2]. Preliminary meetings with women partners helped shift the focus of this project from the communication of e-waste pollution solutions to raising general awareness of pollution. Women working at the Agbogbloshie e-waste site did not mention long-term contamination from e-waste when describing their health concerns. They do not understand the risk associated with exposure to toxic contaminants. This issue requires attention because awareness of contamination is instrumental in preventing its health detriments. Women's duties in food preparation and childcare emphasize the necessity of their lessening the risk of contamination. This project involved collaborating with women living on the Agbogbloshie site to communicate the risks posed by e-waste.

The project focused on understanding the traditional roles of women at the Agbogbloshie e-waste site and co-design with them an awareness presentation of the health risk associated with exposure to toxic substances found in e-waste. This case study aims to present potential health-related problems caused by the exposure to e-waste pollution to help women workers at Agbogbloshie make informed decisions to benefit their long-term health and those of their children. Future work should go be-

Acknowledgments: We thank our advisors, Professor Robert Krueger, Professor Hermine Vedogbeton, and Professor Mustapha Fofana (Worcester Polytechnic Institute), for providing structure and support throughout this project. Their enthusiasm, knowledge, and experience have been vital to its completion. Our sincere thanks to Julian Bennett for serving as our point of contact in Ghana and organizing collaborative meetings with our partners over there. Most importantly, we thank our collaborators in Agbogbloshie, especially those who worked directly with us over the course of the term. Their input and co-design efforts improved this project significantly and are appreciated.

Grace Fitzpatrick-Schmidt, WPI student, Chemical Engineering
Helen Le, WPI student, Computer Science and Robotics
Ethan Wilke, WPI student, Biomedical Engineering
Nicholas Masse, WPI student, Computer Science

https://doi.org/10.1515/9783110786231-005

yond this awareness presentation and work with women partners on workable solutions to mitigate the effect of contaminants and pollutants from e-waste sites.

To convey the process through which this project was co-designed, this chapter will discuss issues associated with e-waste pollution on women's health. First, this chapter describes the site on which our partners live and their routines, then the chapter contextualizes the health risks posed by long-term exposure to an e-waste site. Next, the chapter focuses on methods and materials of exposure and how pollutants affect women and the people around them. This leads to possible solutions to these issues and a summary of the background information. Once the background is established, the methods for the execution of the project are detailed. This involves describing the research that forms much of the background, as well as ideation with partners in Agbogbloshie. The information gathering detailed here is examined and reported in Section 4.8, and the makeup of which will be qualitative regarding the satisfaction of our community partners. This is then discussed in the following section, examining the impacts of the project and how generative justice is achieved. This is important in determining the long-term effectiveness of the project to help people create success for themselves. The conclusion wraps up this chapter and lists possible avenues for further research. This chapter is not a completion of this study: It provides a point from which further projects can begin and apply this research.

4.2 Background and literature review

4.2.1 About Agbogbloshie

Agbogbloshie, Ghana, is one of the top five importers of e-waste in the world and known for hosting an intensive e-waste scrapyard [3]. The e-waste site serves as an informal workplace for many including women. Most women do not work directly with e-waste but operate in close proximity as merchants. They sell food to the scrap dealers, and in the process unknowingly expose themselves to toxic materials [4]. Women are often exposed to air pollution such as smoke from burning plastics or dust-containing toxins. They are also exposed to e-waste toxins through the goods they sell and consume. As a result, these women are equally exposed to contaminants as are the male workers.

The workers on the site are from low-income families and do not have any other means of survival. Many have migrated from Tamale, in the Northern region, where there are fewer job opportunities. "We don't get any jobs," says Yusuf Ibrahim, a 37–year-old scrap dealer who has worked in Agbogbloshie for over a decade [4]. E-waste sites are highly competitive because of the opportunity they provide. Jobs are not plentiful in other environments, so despite the difficult conditions, many lower income families find themselves on the site as they cannot afford to risk relocating and finding employment elsewhere.

4.3 Health risk and contamination sources for women

4.3.1 Sources of e-waste pollution

Common sources of pollutants are found in waste products used for data processing (e.g., computer, keyboard, printers, and scanners), household appliances (e.g., refrigerators), and so on. Batteries and wires are abundant in these modern e-wastes and contribute heavily to deposits of heavy metals and other toxins.

Batteries often contain Pb, Cd, Ni, manganese (Mn), and lithium (Li), all of which are harmful if ingested [32]. Exposure can occur due to groundwater leaching, in which liquid from the batteries containing these metals releases to rivers or is absorbed by the soil in which food crops are grown and on which livestock are raised [3].

Wires and other electronic scraps also threaten the health of residents of e-waste sites because of the plastic they release when burned or stripped. Open burning of printed circuit boards and cables, for example, is commonly used to remove the plastic coating from wires in bulk. This informal e-waste recycling technique, among others, causes the release of contaminants affecting e-waste residents. Not only workers directly involved with waste processing are at risk of exposure, but also food contamination affects all residents of an e-waste site. The pathways to exposure to this contamination must be understood for a solution to be modeled.

4.3.2 Pathways of e-waste pollution

Pollution primarily enters residents' bodies through food, water, and air contamination. Through the air, contaminants commonly reach people by smoke or dust ingestion, often through inhalation. Open air burning of e-waste is proven to affect human health. Heavy metals are detected in adults who reside in e-waste polluted areas in India and are associated with increased prevalence of cardiovascular morbidity, specifically hypertension [5]. This study has application in Ghanaian e-waste sites due to the frequent practice of burning electronics to isolate the valuable metals from the plastics and other less useful components.

Alongside open burning smoke, exposure to the hazardous components of e-waste can occur through inhalation from the air, dietary intake, and soil/dust ingestion [6]. Dust poses a respiratory and dietary risk. It can contaminate the soil where crops are grown and animals are raised, and it can settle on food sold on or near the e-waste site. This dust can contain particles of heavy metals and plastic contamination, which poses long-term health risks similar to smoke inhalation.

Food, in particular, is a dangerous pathway for exposure because the paths previously discussed create pollution to accumulate in the food chain and foodstuffs (i.e., fish from high trophic levels or lipid-rich tissues) [6]. The communal food preparation common on these sites means any problem one batch of food has will be introduced to all who eat from that batch. The numerous pathways to exposure present a challenge to those wishing to curb the health detriments caused by e-waste, forcing one to consider the issue from multiple angles and create solutions that do not interfere with one another or the culture in which they are being implemented. Figure 4.1 shows an example of how dioxins can travel through many sources and end up harming the e-waste workers or inhabitants surrounding an e-waste site. This diagram shows there is both direct and indirect exposure to toxins.

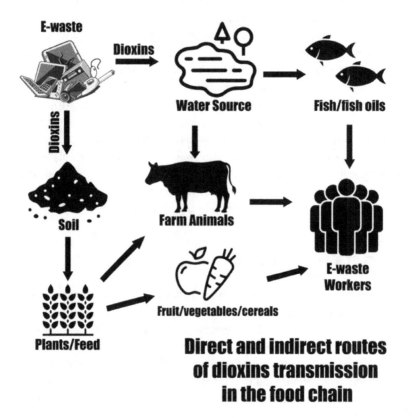

Figure 4.1: Diagram of toxicant pathways from e-waste.

4.3.2.1 Pollutants

Prominent categories of pollutants found on e-waste sites are heavy metals, persistent organic pollutants, and others including dioxins, PCBs, PAHs, and furans. These pollutants come from e-waste sources previously described and have different effects on the human body.

4.3.2.1.1 Heavy metals

Lead is a prominent heavy metal, with increased concentrations in cord blood and meconium correlating with significantly worse behavioral neurological assessment scores among women living in and around e-waste sites compared to elsewhere and could lead to the development of schizophrenia [7]. Lead also has been linked to delayed puberty in girls, and childhood exposures to "polychlorinated biphenyls, lead, mercury, and aluminum have led to changes in mental health-behavioral disturbances, attention deficits, hyperactivity, and conduct issues" [7].

For infants and newborns, "lead in blood was associated with a smaller head circumference, as well as a lower Ponderal Index," which is a measurement used to determine human leanness as the ratio of height to body mass. High cadmium in blood was associated with lower birth weights, BMI, and Ponderal Index [8].

Overall, heavy metals are often associated with increased chances of cardiovascular morbidity, especially hypertension, as shown in research from Indian waste sites [3]. Unfortunately, there is still a huge lack of research on the effect of heavy metals on pregnant women, so there could be many more adverse health effects currently unknown.

4.3.2.1.2 Persistent organic pollutants

Persistent organic pollutants are a class of contaminants that include lipophilic, bioaccumulative substances. These pollutants are extremely resistant to break down because of long half-lives [7], meaning people have a greater chance of encountering them.

These persistent organic pollutants are especially dangerous to unborn children, with many adverse birth outcomes associated with increased exposures to these pollutants, such as polycyclic aromatic hydrocarbons, polybrominated diphenyl ethers, polychlorinated biphenyls, and perfluoroalkyls. Some tragic effects impacting births include spontaneous abortions, stillbirths, and premature births. Those who are born are reported to have birth weights and lengths well below normal.

Additionally, polychlorinated biphenyls, 2,3,7,8-tetrachlorodibenzo-para-dioxin (TCDD), and perfluoroalkyls decrease sperm quality as well as female fertility. These organic pollutants are also endocrine disruptors. They have a much stronger effect especially due to long-term exposure and accumulation – a danger for families and children raised on e-waste sites.

4.3.2.1.3 Other pollutants

For pregnant women, high PCB and PBDE levels were shown to be associated "with thyroid-stimulating hormone and total thyroxine." Abnormal levels of these would cause drastic changes in weight, body temperature, and muscle strength.

All these pollutants contribute to health detriments, especially for pregnant women and young children growing up with constant exposure. They have been associated with "reduced neonatal behavioral neurological scores, increased rates of attention-deficit/hyperactivity disorder, behavioral problems, changes in child temperament, sensory integration difficulties, and reduced cognitive and language score" [33].

Lung and respiratory functions are disrupted, DNA could be damaged, and thyroid functions impaired from contamination. There is also an increased risk of lifelong chronic diseases. "A child who eats just one chicken egg from Agbogbloshie . . . will absorb 220 times the European Food Safety Authority daily limit for intake of chlorinated dioxins" [1]. Throughout a person's lifetime, this exposure to toxins can have serious health effects.

4.3.3 Effects of e-waste pollutants on women's health

E-waste pollutants have unique effects on women due to their constant exposure and close contact with the site: "Women are also particularly at risk to environmental hazards throughout the site" [3]. This stems from a deeper social issue of class with women making up the lower classes. Therefore, the dangerous and undesirable jobs women take directly expose them to the pollutants of the site, and they have little social capital to use in making a change. "E-waste specifically affects women's morbidity/mortality, and fertility, as well as the health of any children" [9]. More than half of the pollutant chemicals found at e-waste sites have a direct effect on women's endocrine and reproductive systems: "12.9 million women are working in the informal waste sector, which potentially exposes them to toxic e-waste and puts them and their unborn children at risk" [34]. In addition to these concerns for women, "18 million children and adolescents, some as young as 5 years of age, are actively engaged in the informal industrial sector, of which waste processing is a sub-sector" (Soaring, 2014). The main concerns for women's health on the e-waste site fall into three main categories: reproductive harm, cancer, and respiratory issues.

4.3.3.1 Reproductive harm

Women working in e-waste sites often encounter fertility problems due to constant exposure to harmful chemicals. Women and their children/fetuses have many unique exposure pathways like "breastfeeding and placental exposures," "high-risk behaviors (e.g., hand-to-mouth activities in early years and high risk-taking behaviors in adolescence), and their changing physiology (e.g., high intakes of air, water, and food, and low rates of toxin elimination)" [7]. Findings indicate there are "consistent effects of exposure with increases in spontaneous abortions, stillbirths, and premature births, and reduced birthweights and birth lengths in most studies" [7]. The fetus is most at risk during the initial stages of pregnancy. "Lead and mercury exposure within the first trimester of pregnancy may affect fetal development, resulting in potential neurobehavioral development problems, low birth weight, or spontaneous abortion and birth defects" [9]. Fetuses are also most vulnerable to dioxin, one of the highly toxic by-products of e-waste. "Newborn[s], with rapidly developing organ systems, may also be more vulnerable to certain effects. Some people or groups of people may be exposed to higher levels of dioxins because of their diet (such as high consumers of fish in certain parts of the world) or their occupation (such as workers in the pulp and paper industry, in incineration plants, and at hazardous waste sites)" [35].

"Despite different exposure settings and toxicants assessed, there have been consistent effects of exposure with increases in spontaneous abortions, stillbirths, and premature births, and reduced birthweights and birth lengths in most studies. Adverse birth outcomes have been associated with increased exposures to polycyclic aromatic hydrocarbons and persistent organic pollutants, including polybrominated diphenyl ethers, polychlorinated biphenyls, and perfluoroalkyls. The main exception to these effects is the lack of association between exposures to metals and adverse birth outcomes" [7].

4.3.3.2 Cancer

In addition to causing reproductive harm, a buildup of dioxins and other pollutants is linked to various forms of cancer. "Dioxins are highly toxic and can cause reproductive and developmental problems, damage the immune system, interfere with hormones and also cause cancer" [10]. "The human body burden of e-waste exposure could cause all kinds of diseases – e.g., cancers, mental health and neurodevelopment disorders, thyroid dysfunction, and general physical health deterioration (DNA damage and effects on gene expression)" [6]. The "persistent organic pollutants [from e-waste] are a group of lipophilic, bioaccumulative substances that are very resistant to breakdown because of long half-lives. Common persistent organic pollutants found in electrical and electronic equipment components include brominated flame retardants (polybrominated diphenyl ethers), polybrominated diphenyls, dibrominated diphenyl ethers, polychlorinated biphenyls, polychlorinated or polybrominated dioxins, and dibenzofurans dioxins, hexabromocyclododecanes, and perfluoroalkyls" [7].

Continued exposure of animals to dioxins has resulted in several types of cancer. TCDD was evaluated by the World Health Organization (WHO) International Agency for Research on Cancer (IARC) in 1997 and 2012. Based on animal data and human epidemiology data, TCDD was classified by IARC as a "known human carcinogen." However, TCDD does not affect the genetic material and there is a level of exposure below which cancer risk would be negligible [7]. "Increase in polycyclic aromatic hydrocarbons (PAHs) pollution in the environment has been of great concerns." These PAHs ". . . are known to pose health problems including various forms of cancers in humans" [11]. These toxins also affect children at an increased rate. "A report by the Cancer Control Division of the Ghana Health Service indicates a rise in the number of cancer cases and young people are in the majority" [11].

4.3.3.3 Respiratory issues

Air pollutants from the e-waste site are linked to a variety of respiratory and lung health concerns in both women and children. "Other adverse child health impacts linked to e-waste include changes in lung function, respiratory and respiratory effects, DNA damage, impaired thyroid function, and increased risk of some chronic diseases later in life, such as cancer and cardiovascular disease" [10]. "Pollutants released during e-waste recycling have been linked to adverse health consequences that include general injuries, respiratory diseases from inhalation, growth retardation, skin disease, immune weakness, neurodevelopmental effects, risks of cancer, increases in spontaneous abortions, and premature births" [12]. "The harmful combustion byproducts released while burning e-waste can increase risk of respiratory and skin diseases, eye infections, and even cancer for people nearby" [13].

4.3.4 Effects of e-waste pollutants on unborn and newborn children

The effect of the pollutants and e-waste on the unborn and newborns are slightly different than those for women. Pregnant women expose their unborn children to pollutants by virtue of sharing a body and newborns through breastfeeding and environmental contamination. There is still a lack of research into pollutants' effects on pregnant women [14]. Children are more vulnerable to pollutants present at the site because of their size, underdeveloped organs, and high growth rate compared to adults on e-waste sites. Any pollutants a pregnant woman is exposed to can cause lifelong health impediments in their children [1].

Heavy metals are commonly found on e-waste sites and can pass through the placenta to the fetus. It has been found that exposure to lead (Pb) and cadmium (Cd) "has been associated with low birth weight and adverse neurodevelopment in children, along with other adverse pregnancy and birth outcomes" [14]. A study conducted at an e-waste site in Guiyu, China, looked at the birth outcomes associated with maternal blood concentrations of different heavy metals. This study compared the concentrations of Pb, chromium (Cr), Cd, and manganese (Mn) in the blood of mothers in Guiyu and a control site, and higher concentrations of Pb, Cr, and Cd were found in the blood from mothers who were from Guiyu, and the Mn concentration was higher for those from the control site. Birth length in Guiyu was also greater than the control site. Birth weight, BMI, Ponderal Index, and head circumference were less in Guiyu, and all were statistically significant. "Guiyu [also] had a higher percentage of preterm birth compared to Haojiang (7.3% vs. 3.0%, respectively), which can cause complications for both mother and child" [14].

This study solely focused on the effects of heavy metals. Exposure to toxins from e-waste has been attributed to stillbirths, premature births, and low birth weight and length. Additionally, exposure to e-waste in children has been linked to "significantly reduced neonatal [behavioral] neurological assessment scores, increased rates of attention-deficit/hyperactivity disorder (ADHD), [behavioral] problems, changes in child temperament, sensory integration difficulties, and reduced cognitive and language scores" [1]. Some toxins from the site are neurotoxicants like "lead, mercury, cadmium, and brominated flame retardants," which are known to cause "cognitive deficits in children and behavioral and motor skill dysfunction across the lifespan" [13].

Babies are primarily exposed to toxins through breast milk, to which there are few alternatives on e-waste sites in Ghana. These pollutants accrue in breast milk through a mother's exposure and are then transferred to the child. The mean concentration of PAHs in breast milk at Agbogbloshie was higher than at the control site, and "the most carcinogenic of all the PAHs, benzo[a]pyrene (BaP), was detected in 92% in the milk samples from Agbogbloshie but were below the limit of detection in all the samples from Kwabenya" [11]. Ingesting carcinogenic PAHs can cause cancer, pulmonary diseases, and mental disabilities [11].

Polybrominated diphenyl ethers found in breast milk were "associated with decreases in the birth outcomes for infants, including birth weight and length, chest circumference" and "CpGs of BAI1 and CTNNA2, which are mostly responsible for neuron differentiation and development, were significantly involved in brain neuronal development in infants and had shown a close association with the elevated maternal Pb levels in the exposed group" [12, 15]. E-waste contamination is so pervasive; women and infants do not have to be near sites to experience adverse effects (Figure 4.2) [16].

4.4 Potential solutions to e-waste contamination

Solutions to e-waste contamination often include food preparation changes, limiting air pollution exposure, water filters, and plants; however, the feasibility of these solutions is largely undocumented. Most pollutants are found in food products, including meat, fish, fruits, and vegetables. Dioxins are found in the fat of meat and dairy products, so limiting fat ingestion can limit exposure. Alternative cooking processes might help remove different pollutants as well. Air pollution is omnipresent on an e-waste site, so "closed windows, usually associated with the use of air conditioning in the developed world, can reduce air exchange rates by about 50%, leading to reduced infiltration of ambient air pollutants to the indoor environment" [17].

Air filtration can reduce the concentration of pollutants as well. Breathing through one's nose instead of their mouth can also reduce ingestion of airborne pollutants: "compared to the mouth, the nose is a more effective filter for preventing particles and water-soluble [gasses] and vapors from reaching the lungs" [17]. Personal protection equipment like respirators can also limit exposure to pollutants in the air [17]. Like with air, filtration is a highly effective method of reducing exposure to toxins from

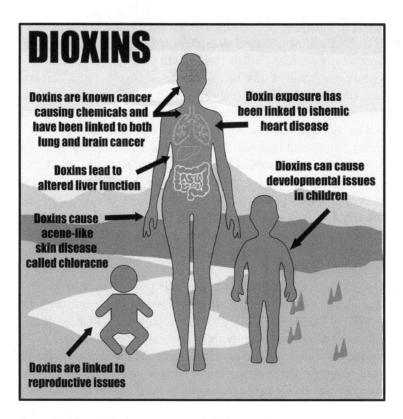

Figure 4.2: Effects of dioxins on women and children.

water or drinking prepackaged water from outside the site. Planting trees and other plants in the area achieves a similar effect, especially with metal-accumulating plants that undergo phytoremediation, treating contaminated soil and water [18]. These methods all prove helpful in theory but may not be plausible in practice when introduced to an e-waste site. Resources are limited and schedules are strict, so many of these tactics may be forgone for preestablished concerns.

Addressing health concerns and communicating risks associated with sub-Saharan Africa are best achieved with a smart village. Educating women on the health risks, they and their children face is important to a smart village. A smart village facilitates the spread of information within a community without drastically altering the livelihoods of those involved. Health issues are communicated to the population, allowing them to decide the path they travel. Creating a smart village must be done using co-design, or it will not be smart for its inhabitants. It must involve providing information to members of the community, so they can decide what must be done based on their knowledge.

4.5 Health risk perception and communication

4.5.1 Perception

E-waste site workers and inhabitants have little concern for the risks posed by e-waste. Amugsi et al. [19] looked at levels of perceived health risks at solid waste dump-sites, where solid waste management (SWM) is examined. Only 42% of the participants at Mombasa and 27% at Nairobi in Kenya "perceived that there was little, or no risk associated with poor SWM. Many of the study participants believed that they were exposed to this risk through bad smell" [19]. Both sites reported about 80% of participants claiming no adverse health effects from the waste. In Nigeria, 88% of the e-waste workers "were unable to mention at least one chemical present in e-waste and were unaware that e-waste contains hazardous chemicals which could harm their health" [20]. The population also showed overall poor knowledge of occupational health risks. This could result from the nature of the risks posed, since many people may not connect long-term health issues with e-waste. Most concerns of the e-waste workers and the people living on the site are with short-term aches and less about long-term problems [19]. The occupational health risk awareness level is dependent on job designation, location, and position in the business, especially among dismantlers [20]. Many workers do not even believe that they could get sick from their job, and that it comes from outside factors [20].

4.5.2 Communication

Communication of health risks is important to help people understand the daily risks they face and how to avoid them. It must be well prepared to ensure effectiveness [21]. Communication has focused on health professionals instead of community members in the past, which fails to alleviate exposure [22]. This means "a two-way interactive process [is necessary]. Merely presenting information without having regard for communicating the complexities and uncertainties of risk does not ensure effective risk communication" [23]. To communicate risk effectively, one should use clear and simple facts with supporting evidence and establish a bottom line immediately [22, 24]. The bottom line ensures that the information conveys a message that is easily understood [24]. Risks should be framed and presented based on the community, so the information feels relevant [21]. Medical professionals are not the best for educating people because they are mistrusted by many Ghanaians. Patients' wariness even leads them to consult the internet instead of doctors for medical advice [25]. When health professionals use inaccessible language to communicate risks, they alienate people even more. Multiple forms of information should be presented, and questions should be answered to encourage understanding [26].

Visual aids show risks without relying on words or numbers, "[improving] risk communication for people with limited language skills and limited medical knowledge" [27]. These pictorial aids can also help patients feel more comfortable trusting physicians than verbal education, and people who benefit from the visual representations are often those more vulnerable to risk. "As long as vulnerable people have moderate levels of graph literacy, appropriate visual aids tend to dramatically improve comprehension and decision making," including bar charts, pie charts, and pictographs [27]. If the level of graph literacy is low, then these visual aids are not as helpful and other methods should be considered.

Brochures and fact sheets efficiently show risks and deliver information; however, they are not particularly effective. One study indicated a fact sheet showed 71% of the participants understood a way to reduce their risk and "booklets worked better than the fact sheets, but no single format appeared to be best" [28]. A classroom presentation proved more effective than either with 96% of participants understanding a risk reduction method because questions can be answered in real time [28]. When many people understand the risks they face, word of mouth becomes the most effective form of information transference [26]. When people hear information from their peers, they are much more likely to trust it than from health or government officials [23].

Media, such as newspapers, magazines, radio, TV, or social media, is another effective way to convey information. Mass multimedia approaches can have a strong influence on risk perception, so they must be well researched [28]. "Both the news media and the Internet have been criticized as often publishing inaccurate, sensationalized, or misleading stories that are not necessarily the most scientifically significant" [21]. The public becomes suspicious of these resources when this happens, and in many instances, this is justified when reporters report poorly prepared information [29]. The variety of languages in Ghana makes media campaigns difficult, but mass multimedia campaigns are the best way to spread information [30]. In sub-Saharan Africa, the most powerful media type is the radio, as film and other modern technologies have failed to communicate reliable information to enough people [29, 30].

Health professionals in Ghana must revere the culture they communicate with. The language must be understandable for the community [28]. Sometimes the direct translations do not convey the correct meaning [22]. Translations should involve community members to ensure understandability and provide more ownership over the project. "The principles of 'inclusion, participation, and self-determination' help defeat the major problems seen with solely increasing comprehension of why a certain health behavior is wrong" [30]. Integrating the community can have benefits that extend just what risks are being communicated.

4.6 Summary

E-waste sites have a wide variety of tasks performed on numerous types of electronics like computers, phones, televisions, and car parts and now provide a livelihood for lower class families in becoming an essential part of the economy [4]. These sites can release pollutants coming from materials like batteries and wires [6]. Rain can exacerbate the pollution, causing runoff that distributes pollutants to the ground and water supply [3]. The five major pathways to exposure are air, water, soil, dust, and food. The most prominent toxins are heavy metals and dioxins, which can cause cancer, neurological problems, thyroid problems, and other health conditions [6, 36]. Toxins can accrue in breast milk and expose infants, so a woman's exposure affects her and her children [11]. Exposure to these toxins can lead to birth defects, low birth weight, stillbirths, preterm birth, and mental issues or developmental delays in children [1]. Some solutions have been explored to help reduce the exposure, including water filters, plants that absorb contaminants in soil, and food preparation techniques like trimming the fat [17, 18, 31]. Many e-waste workers and residents do not perceive these health risks as a threat [19]. Solutions like visual aids or lessons must be co-designed to help people without disturbing their culture or livelihood.

4.7 Methods and materials

4.7.1 Research

Due to the nature of the project and how little was known about Ghana and e-waste sites, the first step of the project was to conduct extensive research into e-waste contamination in Ghana and how it affects the people. The research was divided into three main research categories/objectives: identifying health risks and contamination sources for women, identifying health risks and contamination sources for unborns/newborns, and how solutions will be implemented. Shared documentation locations were made for each objective and over the course of 2 weeks, the project members used online journal databases and search engines to find and report information into the documents. The project members met three times a week to discuss the individual research with each other to set proper objectives for the next meeting, ensuring the research was always moving forward. We increase our knowledge on potential solution to e-waste pollution by studying e-waste projects by Harvard School of Engineering and Applied Sciences (SEAS), WHO, and the Ghana Health Service. The variety of background information helped create better situational awareness of what is going on in the e-waste site regarding contamination and day-to-day life.

The other main area of research revolved around the basics of the e-waste site, who lives there, what they eat, where they sleep, and so on. This research was conducted to provide the project members with a better understanding of what life on the e-waste site is, specifically the one in Agbogbloshie. Most of this research was conducted using online videos and articles provided by the residents of the e-waste sites, in addition to various drone and aerial photography provided by the advisor and Google Maps. This research provided the members an inside look into what occurs in the day-to-day life on the site, which was effectively reported through the text of the previously read journals. While a document was created and completed using the information found, this research was more importantly used subconsciously throughout the meeting and design processes to make the project members more efficient at co-design with our Ghanian partners.

Once the initial research phase was complete, the documents were examined by all members to ensure a unanimous understanding of the information. The material was then organized using the outline of the case study as a reference, using its different subcategories to organize the research. Now in order, the research was directly used in the background and introduction sections of the case study, where the information could be easily found and referenced. The database of the research was constantly referenced throughout the meeting, design, and implication steps of our project when needed.

One challenge our project team had to work through was defining and creating the scope for our solution. The overall goal of working with the Ghanaian partners to reduce the exposure of e-waste contaminations for women and children was a constant, but how that would be accomplished shifted a few times throughout the project. To organize our thoughts and ideas, we created several charts and documents to help visualize the next steps and the feasibility of an idea. A schematic was created (Figure 4.3) to provide the group members a better outlook on the outline and any changes to make as the project moves toward the final stages.

Once the scope of the project was altered, as shown in Figure 4.3, two additional research categories emerged. The first is how health risks are perceived on e-waste sites and in sub-Saharan Africa in general and the second is how health risks or risks, in general, are communicated to the public. This research took the same process of looking at journal databases and government sites to find sources and articles relating to those topics. Then these sources were used to develop more background and lead to evaluating the best practices for communicating health risks on the Ghana e-waste site to women about themselves and their children. The research was evaluated by organizing and narrowing down the most effective ways to communicate to the women taking into context where they live and their culture. This was the initial evaluation completed and then taken to our meetings with our community partners to be further discussed.

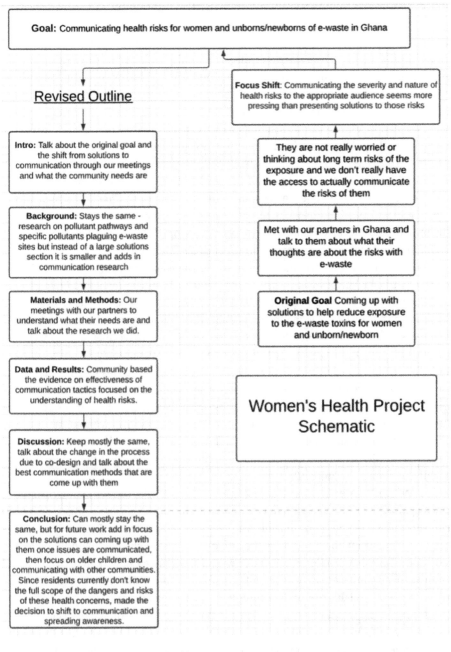

Figure 4.3: Schematic of the women's health project.

4.7.2 Ideation sessions with community partners

To create the groundwork of an effective project, co-design was used throughout the research and design of our solution. Our community partners in Ghana were an essential aspect of our project as they collaborated with us during the research and design phase but more importantly would continue our project into the future to make sure its impact extends beyond our brief time of influence. Due to the social environment created by COVID-19, our meetings with the project partners in Ghana were conducted remotely. This made implementing the co-design plan more difficult but still possible. To make the best of the virtual meetings, time was spent beforehand familiarizing the team with the daily life of people on the e-waste site so that information could be compared, and questions could be quickly generated. The initial plan was to create a list of meeting questions but after deliberation and some time and thought, the project team decided it more beneficial to develop a list of objectives/topics to talk about to make conversations more natural and less interview-like.

The first meeting was conducted over Zoom. The meeting was with a female Agbogbloshie worker who sells water sachets (Respondent 1). Julian Bennett, a Ph.D. student at Academic City University in Ghana, brought a laptop to the site to serve as our ambassador and translator for the meeting. This was a good initial meeting and gave the members a decent reference for what to expect from future meetings. The group learned a lot about the worker from this session, including her daily diet, what she did for a living, and thoughts on contamination on the e-waste site. This meeting had a considerable influence on our project scope as we learned that people on the site not only did not consume a large amount of meat but also had no personal experience with health concerns from e-waste contamination. This led us to believe that the workers and women on the site might not understand or know about the long-term health consequences of the e-waste. After this meeting, the goal and focus of the project shifted slightly, as shown in Figure 4.3.

A few weeks later another meeting was conducted with our community partners. Once again Julian Bennett served as an ambassador and translator. This meeting included another female Agbogbloshie worker who sells bananas (Respondent 2). The group learned from about her daily life and diet, concerns she has about living on the site, and where her food comes from. It was also discussed how their routines could change to minimize risks of exposure to contaminants. Different communication styles to present health risk to the community have been also discussed.

A connection was made with students at the University of Ghana in Accra, Ghana, who are interested in working on this project. Two meetings were held with the student partners to discuss the project and their ideas. The first meeting included introductions and discussions about what was accomplished so far with the project and what their initial ideas were about it. The second meeting was after the students were able to travel to the site. This meeting gave a better insight into life on the site, perception of health risks, and how to best communicate to the community.

A meeting was also held with environmental photojournalist Mike Anane. He spent time on the site taking photos of the people living and working there. He also traveled to other smaller e-waste sites in Ghana, so he is experienced with life on sites on a smaller scale than Agbogbloshie. During conversations with Mike, he gave us insights into the lives of people on the site, what they have access to, and how things are communicated. This conversion helped to reinforce ideas we held, although it sometime contradicted the information we received from others.

4.7.3 Development of graphics

The last part of the project was developing graphics based on our research and conversations. Five different graphics were created. The first showed where and how toxins affect women, children, and infants. The second presented the traveling of dioxins through the ecosystem and how they find their pathways to humans. The third displayed different pathways, sources, and health effects for different pollutants. The fourth presented the pros and cons of different communication styles done in the past. And the fifth illustrates the communication styles chosen as the best for the community and highlights how they can be used to communicate the health risks of the e-waste.

4.8 Data and results

Respondent 1 says she typically gets to the site around 7:00 am and starts setting up her station to sell the water sachets. She works until around 3:00 pm when she takes a break and then typically works to 5:00 pm to 6:00 pm depending on how business is. She usually comes to work every day. Respondent 2 lives close to the site and gets up around 5:00 am and walks 5 km away from the site to buy bananas to sell. Around 11:00 am, she typically starts selling, walking all around the site. She will end business between 5 pm and 8 pm depending on how commerce is that day.

Our women community partners at the site said that almost all their food comes from the site or from farms nearby, which means this is where they grow it and buy it. Breakfast for these women is typically tea and porridge. For other meals, they have waakye, which is rice and beans typically eaten with stew and eggs. Another dish is T-zed, which is a northern bean dish. They also eat fish from the nearby river. Their water for drinking purposes is sold in sachets called Good Pack. The water from the tap is not consumed; it is just used for cleaning. The food is typically prepared by a group of 20 women for about 1,000 people on the e-waste site. This food made by the women is sold to the workers. To cook they use cobots, which are communal fire/gas stoves. There are also several types of fruits and vegetables as a part of their diet, though specifics of what type of fruits and vegetables are available were not mentioned, other than bananas.

There is not much about women's daily routines/lives that can be changed. The goal on the site is to make money, and other things are not of great concern to them. Respondent 2 said that if daily routines were to change, it would be easier to change that of the men working on the site. The methods for cooking are fairly permanent because they do not have the resources to adapt to new methods. From the conversation with Respondent 2, she said that any type of communication is good, and it should be in the local language of Twi because not everyone knows English.

Conversations with Mike Anane revealed young children (usually boys) as young as 7-year-old dismantle e-waste without personal protective equipment. Men are the dominant demographic in physical e-waste processing, but women assist with burning and some disassembly. Mike also believes most people are unaware of the toxicity of e-waste and its detrimental impacts. Those aware know because of personal experience. This reinforces the idea that communication of risk is the most important next step for avoiding the health risks of e-waste.

Overall, from the conversations with our community partners, there is a low-risk perception about the e-waste and the toxins that the e-waste contains or produces. They said that some health issues they saw from the e-waste were occasional sickness, headaches, and other bodily pain from carrying water and daily strain, fever, and stomachache. Some concerns they reported were they got lots of cuts, back pain, eye irritation, coughing, and encountered rain bringing waste into the area and mosquitoes. These are just overall concerns and not specific things relating to e-waste. With children, they just see basic cold symptoms and are not concerned much about children's health on the site. There are some people who are aware of the risks but accept them as part of daily life. Their focus is to make money, which means that things potentially preventing that from happening are ignored.

4.9 Discussion

4.9.1 How to communicate

After conversations with our Agbogbloshie partners, the project's focus became communication of health risks to the people living or working on the site. From research and the conversations, best practices for communication were determined. These can be implemented in the future to be communicated to the women living and working on the site. Community integration is important for the communication of these risks. The community should be involved in creating teaching materials and communicating. The translation is another major responsibility that must fall on community members. This promotes greater ownership of ideas and motivation for others to listen while bridging the language gap. Agbogbloshie student partner meetings revealed many people at the site do not speak English and the variety of dialects makes communication difficult. Community leaders are ideal candidates for translation due to their high standing and outreach within their communities. This means providing leaders information and allowing them to translate and distribute it. Visual aids are a helpful supplement to verbal communication in distributing health information. It is important for graphics to limit words to extend accessibility. These aids are likely to be thrown out if distributed by foreigners, so community leadership is important in this as well. Technology-based communication is not the best way to communicate on Ghanaian e-waste sites. Residents often have limited education, especially regarding technology.

Figure 4.4 displays the pros and cons of communication methods regarding e-waste communities. It is important to consider the details of each method to determine the most effective one.

Pro/Cons of Communication

	Word-of-mouth/ In person Presentations	Media (Electronic)	Media (Print)	Brochures/ Fact sheet
Communication Style	Classroom Presentation Town/Village Meetings Talking in Small groups Doctor One on One	TV Radio Internet Social media Phone calls	Newspapers Magazines	Booklets Fact sheet Brochures Pamphlets Posters
Types	PROS -Typically done by experts -People can ask questions/clarify -More personal -Community integration -When done in some language takes that barrier away	PROS -Quick + widespread -Easily translated -Can have graphics	PROS -Easily distributed	PROS -Usually have graphics to share things other than numbers -More visually appealing
	CONS -Could get too technical or be in a different language -More time consuming -Requires people to be there -Only really shows numbers	CONS -Not everyone has access -People don't answer their phones -Not always from experts	CONS -Typically one language -Expensive -Not always from experts	CONS -Expensive -Needs to be translated

Figure 4.4: Pros and cons of different communication methods.

Figure 4.5 shows a guide on how the health risk information can be communicated in an effective manner to the women on the site. This can be used as a guide for future projects on how to communicate the health risks of e-waste to women so they can help them implement ways to reduce their exposure to e-waste.

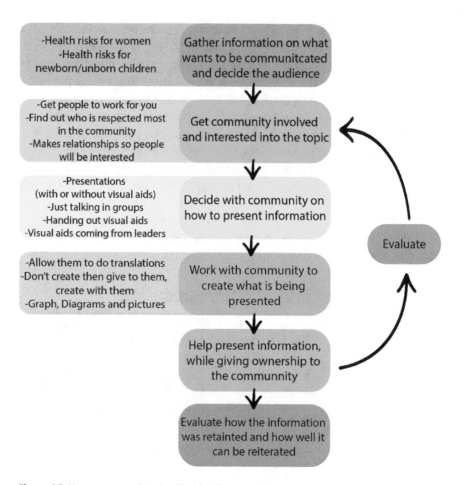

Figure 4.5: How to communicate health risks of e-waste in Ghana.

4.9.2 Working alongside Agbogbloshie

The opinions of Agbogbloshie residents are vital to the core of this project. In meeting with community partners, this project shifted focus from solution ideation to communication of risks. This is because many women are not aware of the health impacts posed by e-waste, and routines are difficult to change on the site. This shift addresses a more prevalent issue, allowing women in Agbogbloshie to change their own lives given the appropriate information. However, the scope of the information gathered has been limited by few meetings with women. The information gathered is useful, albeit limited in scope by logistical issues. These conclusions were derived from discussion with these women and consideration of their values and needs.

Generative justice is crucial to our interactions with Agbogbloshie's residents. Our meetings with them were limited, so we implemented generative justice through questions about the women's daily lives, interactions with the site, and food preparation and sale. These questions incorporate generative justice by accounting for women's issues. The project is built around women because women should take charge of what they have created with us. Initially, the project plan was to design and implement solutions for women to mitigate the effects of the contaminants on the e-waste site, but they are not aware of the toxic effects of living on an e-waste site. The project instead focuses on communicating the health risks of e-waste to Agbogbloshie's residents. This includes providing visual aids and graphics in Twi, as well as giving this information to leaders, lending credibility to the information provided.

The intended result of the collaboration is to work with the women of Ghana to solve problems affecting women and children. Some unnecessary exposure to contaminants may be limited, and the corresponding health risks to the women and children will follow. Women on e-waste sites will therefore have a course of action to beneficially change their lives and those of all members of their community.

4.10 Conclusion

E-waste sites harm the health of all who inhabit them, severely impacting women and infants who are often overlooked due to their mostly auxiliary role on the sites. Selling food and water near a site can provide as much risk as working directly with the waste, but to improve long-term health, they must understand the risks of their environment. Women often do not understand the impact e-waste and its toxins have on them. Communicating the risks associated with e-waste living is the next step in ensuring residents lead healthier lives.

Given more time, further research would explore the impact of an e-waste site on older children (toddlers, school-aged children, and adolescents). Early childhood development contains the most drastic change for the human body, so e-waste contamination is especially harmful in this period. Pathways like a children's play area should be considered in future studies as well. This could be solved by moving their play area away from most e-waste. Animals and other food could be moved from the site to help mitigate dangerous toxic exposure. In Agbogbloshie, food is handled near waste, and animals roam the site. A new site brings new opportunities to alter methods for living, like allowing animals to roam in an area cordoned away from e-waste. Once the issues are properly communicated and understood, they can be solved with the cooperation of Teacher Mante residents.

References

[1] Johnson, C. (2021). Soaring E-Waste Affects the Health of Millions of Children, Who Warns. World Health Organization. https://www.who.int/news/item/15-06-2021-soaring-e-waste-affects-the-health-of-millions-of-children-who-warns.

[2] Weber, E. U. (2006). Experience-Based and Description-Based Perceptions of Long-Term Risk: Why Global Warming does not Scare us (Yet). *Climatic Change, 77*, 103–120. https://doi.org/10.1007/s10584-006-9060-3.

[3] Sovacool, B. K. (2019, October 4). Toxic transitions in the lifecycle externalities of a digital society: The complex afterlives of electronic waste in Ghana. Resources Policy. Retrieved February 3, 2022, from https://www.sciencedirect.com/science/article/pii/S0301420719302053?casa_token=2eer6a P8LoMAAAAA%3ARUdXNjeU0tCinPmtF8gV74r8h1f2DvjlFd3L24nEXUlCNM3h8ILfD4NlbrgauGz5jy393XKa

[4] Lawal, S. (2019, February 14). Women Living at Accra's Toxic Scrapyard. Women's Media Center. https://womensmediacenter.com/news-features/women-living-at-accras-toxic-scrapyard.

[5] Gangwar, C., Choudhari, R., Chauhan, A., Kumar, A., Singh, A., & Tripathi, A. (2019). Assessment of Air Pollution caused by Illegal E-waste Burning to Evaluate the Human Health Risk. *Environment International, 125*, 191–199.

[6] Song, Q. & Jinhui, L. (2014). A Systematic Review of the Human Body Burden of E-waste Exposure in China. *Environment International, 68*, 82–93. Retrieved February 3, 2022, from https://www.sciencedirect.com/science/article/pii/S0160412014000919?casa_token=nwA6FA-uKkEAAAAA%3AOiv_7ZGkW9W8LDDd6kT9eqdl_Qzgf97wJvy2opbQM-oB6wGapsoz2UmVek24nu3saE0WwaCc.

[7] Grant, K., Goldizen, F. C., Sly, P. D., Brune, M.-N., Neira, M., Van den berg, M., & Norman, R. E. (2013). Health Consequences of Exposure to e-Waste: A Systematic Review. *The Lancet Global Health, 1*(6), 350–361. https://doi.org/10.1016/s2214-109x(13)70101-3.

[8] Shih, Y.-H., Yun Chen, H., Christensen, K., Handler, A., Turyk, M. E., & Argos, M. (2021). Prenatal Exposure to Multiple Metals and Birth Outcomes: An Observational Study Within the National Children's Study Cohort. *Environment International, 147*.

[9] McAllister, L., Magee, A., & Hale, B. (2014). Women, e-waste, and technological solutions to climate change. *Health and Human Rights, 16*, 166.

[10] World Health Organization. Soaring E-waste affects the health of millions of children, WHO warns. (2022). Retrieved February 3, 2022, from https://www.who.int/news/item/15-06-2021-soaring-e-waste-affects-the-health-of-millions-of-children-who-warns

[11] Asamoah, A., Nikbakht Fini, M., Kofi Essumang, D., Muff, J., & Gydesen Søgaard, E. (2019). PAHs Contamination Levels in the Breast Milk of Ghanaian Women from an e-Waste Recycling Site and a Residential Area. *Science of the Total Environment, 666*, 347–354. Retrieved February 3, 2022, from https://doi.org/10.1016/j.scitotenv.2019.02.204.

[12] Singh, N., Ogunseitan, O. A., & Tang, Y. (2020 July 9). Systematic Review of Pregnancy and Neonatal Health Outcomes Associated with Exposure to e-Waste Disposal. *Critical Reviews in Environmental Science and Technology, 51*(20), 2424–2448. https://doi.org/10.1080/10643389.2020.1788913.

[13] Heacock, M., Bain Kelly, C., Ansong Asante, K., Birnbaum, L. S., Lennart Bergman, Å., Bruné, M.-N., Buka, I., et al. (2016). E-waste and Harm to Vulnerable Populations: A Growing Global Problem. *Environmental Health Perspectives, 124*(5), 550–555. Retrieved February 3, 2022, from https://ehp.niehs.nih.gov/doi/full/10.1289/ehp.1509699.

[14] Kim, S. S., Xijin, X., Zhang, Y., Zheng, X., Liu, R., Dietrich, K. N., Reponen, T., et al. (2020 April). Birth Outcomes Associated with Maternal Exposure to Metals from Informal Electronic Waste Recycling in Guiyu, China. *Environment International, 137*, 105580. https://doi.org/10.1016/j.envint.2020.105580.

[15] Xinghong, L., Tian, Y., Zhang, Y., Ben, Y., & Quanxia, L. (2017). Accumulation of Polybrominated Diphenyl Ethers in Breast Milk of Women from an e-Waste Recycling Center in China. *Journal of Environmental Sciences, 52*, 305–313. https://doi.org/10.1016/j.jes.2016.10.008.

[16] Kim, S., Xijin, X., Zhang, Y., Zheng, X., Liu, R., Dietrich, K., Reponen, T., et al. (2018, August 15). Metal Concentrations in Pregnant Women and Neonates from Informal Electronic Waste Recycling. *Journal of Exposure Science & Environmental Epidemiology, 29*(3), 406–415. https://doi.org/10.1038/s41370-018-0054-9.

[17] Laumbach, R., Meng, Q., & Kipen, H. (2015, January). What Can Individuals Do to Reduce Personal Health Risks from Air Pollution? *Journal of Thoracic Disease, 7*(1), 96–107. https://doi.org/10.3978/j.issn.2072-1439.2014.12.21.

[18] Qin, G., Niu, Z., Jiangdong, Y., Zhuohan, L., Jiaoyang, M., & Xiang, P. (2021). Soil Heavy Metal Pollution and Food Safety in China: Effects, Sources and Removing Technology. *Chemosphere, 267*, 129205. https://doi.org/10.1016/j.chemosphere.2020.129205.

[19] Amugsi, D. A., Nigatu Haregu, T., & Mberu, B. U. (2019, April 8). Levels and Determinants of Perceived Health Risk from Solid Wastes among Communities Living near to Dumpsites in Kenya. *International Journal of Environmental Health Research, 30*(4), 409–420. https://doi.org/10.1080/09603123.2019.1597834.

[20] Ohajinwa, C. M., Van Bodegom, P. M., Vijver, M. G., & Peijnenburg, W. J. G. M. (2017, August 13). Health risks awareness of electronic waste workers in the informal sector in Nigeria. In *MDPI*. Multidisciplinary Digital Publishing Institute. https://doi.org/10.3390/ijerph14080911.

[21] Glik, D. C. (2007, April 21). Risk Communication for Public Health Emergencies. *Annual Review of Public Health, 28*(1), 33–54. https://doi.org/10.1146/annurev.publhealth.28.021406.144123.

[22] Finau, S. A. (2000 July 1). Communicating Health Risks in the Pacific: Scientific Construct and Cultural Reality. *Asia Pacific Journal of Public Health, 12*(2), 90–97. https://doi.org/10.1177/101053950001200207.

[23] Nicholson, P. J. (1999). Communicating Health Risk. *Occupational Medicine, 49*(4), 253–256. https://doi.org/10.1093/occmed/49.4.253.

[24] Wilhelms, E. & Reyna, V. (2013 January 1). Effective Ways to Communicate Risk and Benefit. *AMA Journal of Ethics, 15*(1), 34–41. https://doi.org/10.1001/virtualmentor.2013.15.1.stas1-1301.

[25] Ancker, J. S., Chan, C., & Kukafka, R. (2009, August 4). Interactive Graphics for Expressing Health Risks: Development and Qualitative Evaluation. *Journal of Health Communication, 14*(5), 461–475. https://doi.org/10.1080/10810730903032960.

[26] Larsson, L. S., Butterfield, P., Christopher, S., & Hill, W. (2006, March 1). Rural Community Leaders' Perceptions of Environmental Health Risks. *AAOHN Journal, 54*(3), 105–112. https://doi.org/10.1177/216507990605400303.

[27] Rocio, G.-R. & Cokely, E. T. (2013, September 25). Communicating Health Risks with Visual Aids. *Current Directions in Psychological Science, 22*(5), 392–399. https://doi.org/10.1177/0963721413491570.

[28] Fitzpatrick-Lewis, D., Yost, J., Ciliska, D., & Krishnaratne, S. (2010). Communication about Environmental Health Risks: A Systematic Review. *Environmental Health, 9*.

[29] Thompson, E. E. (2019, September 27). Communicating a Health Risk/Crisis: Exploring the Experiences of Journalists Covering a Proximate Epidemic. *Science Communication, 41*(6), 707–731. https://doi.org/10.1177/1075547019878875.

[30] Prilutski, M. A. (2010). A Brief Look at Effective Health Communication Strategies in Ghana. *The Elon Journal of Undergraduate Research in Communications, 1*(2), 51–58.

[31] World Health Organization. Dioxins and their effects on human health. *Media Centre Fact Sheet* (2010). Retrieved February 3, 2022, from https://www.who.int/news-room/fact-sheets/detail/dioxins-and-their-effects-on-human-health

[32] Kim, T., Song, W., Son, D. Y., Ono, L. K., & Qi, Y. (2019). Lithium-ion Batteries: Outlook on Present, Future, and Hybridized Technologies. Journal of Materials Chemistry A, 7(7), 2942–2964.

[33] Chung, E. H., Chou, J., & Brown, K. A. (2020). Neurodevelopmental Outcomes of Preterm Infants: A Recent Literature Review. Translational Pediatrics, 9(Suppl 1), S3.

[34] World Health Organization. (n.d.). Soaring E-waste Affects the Health of Millions of Children, Who Warns. World Health Organization. Retrieved February 3, 2022, from https://www.who.int/news/item/15-06-2021-soaring-e-waste-affects-the-health-of-millions-of-children-who-warns.

[35] World Health Organization. (n.d.). Dioxins and their Effects on Human Health. World Health Organization. Retrieved February 3, 2022, from https://www.who.int/news-room/fact-sheets/detail/dioxins-and-their-effects-on-human-health.

[36] Parvez, S. M., Jahan, F., Brune, M. N., Gorman, J. F., Rahman, M. J., Carpenter, D., Islam, Z., Rahman, M., Aich, N., Knibbs, L. D., & Sly PD. (2021 Dec). Health Consequences of Exposure to E-waste: An Updated Systematic Review. Lancet Planet Health, 5(12), e905–e920. doi: 10.1016/S2542-5196(21)00263-1. PMID: 34895498; PMCID: PMC8674120.

Marika Bogdanovich, Julia Leshka Jankowski, Elizabeth DiRuzza,
Sophie Kurdziel, Suali, Julian Bennett, Melinda Tatenda,
Alice Abigail Tatenda Bere, Farouk Tetty-Larbie

Chapter 5
Stirling engine design using recycled materials locally in Ghana

5.1 Introduction

Ghanaian citizens face unreliable and unaffordable electricity in their everyday lives that many in the Western world take for granted. The team sought to find a cheap, sustainable, and reliable source of electricity for charging cell phones in remote places of Ghana. A Stirling engine is used to harvest the current–voltage supply required to charge cell phones in Ghana. The Stirling engine assembly includes a flywheel, piston–crank mechanism, pulley, belt, bearings, servomotor, and a thermoheat source flame. The performance of the Stirling engine is evaluated to possibly replace the engine components with e-waste materials available in Ghana. Experimental testing, modeling, component design engineering, and analysis were carried out to locate conditions for effectively generating the current–voltage supply for charging cell phones in Ghana. A current–voltage power supply of 4.8 V was generated by the Stirling engine. An $R_aC_aL_a$ electric circuit for the servomotor in the Stirling engine assembly, which is analogous to the mechanical mass–dashpot–spring system, was used to establish the necessary conditions for effectively producing the required voltage for charging cell phones in Ghana's remote places. The engineering design process, experimental work, and analysis form the basis for the creation of guidelines for selecting and replacing the Stirling engine components with e-waste materials. Engaging with our Ghanian partners and the co-design activities highlight the broad impact of our work. The established performance conditions also guide the selection and replacement of the Stirling engine components with available e-waste materials in Ghana.

Marika Bogdanovich, WPI student, Mechanical Engineering
Julia Leshka Jankowski, WPI student, Mechanical Engineering and Environmental and Sustainability Studies
Elizabeth DiRuzza, Mechanical Engineering
Sophie Kurdziel, WPI student, Data Science
Suali, E-waste processor and computer repairer, Agbogbloshie, Ghana
Julian Bennett, Lecturer, Academic City University College, Ghana
Melinda Tatenda, Alice Abigail Tatenda Bere, Farouk Tetty-Larbie, ACUC student, Mechanical Engineering

https://doi.org/10.1515/9783110786231-006

We measured the amount of energy transferred as work to the components of the Stirling engine over some infinitesimal displacement in terms of speed for producing the current–voltage supply into the charging system. A workable prototype of a Stirling engine that could be created using local materials was provided to the students and professors at Academic City University College (ACUC) in Ghana. The local materials include what is available at the e-waste site in Accra, Ghana. Using Ghana's local materials would allow Worcester Polytechnic Institute (WPI) and ACUC students to extend the derived results in this Major Qualifying Project (MQP) and to create a multitude of iterations of the new design.

5.1.1 Uneven electricity access in sub-Saharan Africa

For the average person anywhere around the world, every aspect of life involves electricity in some form. Access to reliable, cheap electricity is something that is taken for granted by people who live in countries with 100% access to electricity. In 2016, 87% of the world's population had access to electricity [1]; however, there are still disparities among countries. As of 2019, most countries have at least 90% access to electricity, as shown in Figure 5.1.

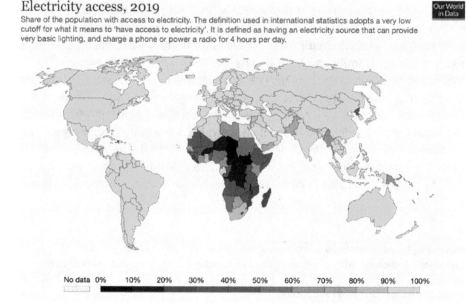

Figure 5.1: Map of electricity access around the world in 2019 [1].

Access to electricity remains a challenging problem in many parts of the world. In Africa, most of the countries have electricity coverage of 60% or less. In some African countries, such as Chad, only 8.4% of the population had access to electricity as of 2019. This is a major issue that our project attempted to address. Ghana has a unique electricity situation since most of its citizens are on a grid system. Despite this, the country experiences issues with reliability due to frequent blackouts. Sudden blackouts are named dumsors, by the Ghana residents, and are a normal part of Ghanaian life with their daily occurrences [2]. Additionally, the cost of electricity is expensive compared to the average income of Ghanaian people. These electricity issues can create major problems for people who rely on electronic products, such as their cell phones. To help minimize this problem, we worked to co-design a Stirling engine out of electronic waste or e-waste that can locally be manufactured and repaired. In total, the cost of materials and labor needed to replicate this project would be lower in comparison to the price of electricity.

5.1.2 Objectives

The project was intended to alleviate these issues while incorporating generative justice and co-design practices. The main goal of this project was to work toward a design of a Stirling engine using e-waste and recycled materials for the purpose of charging a cell phone. This was accomplished through four objectives:
- investigate and reverse engineer a Stirling engine to understand how the engine functions to later alter the material design;
- develop a proof of concept to better understand the basic function and purpose of the Stirling engine using calculations and testing;
- identify and evaluate e-waste materials to clarify which parts could have the potential to work on the Stirling engine; and
- substitute or replace components in a functioning Stirling engine using materials collected from e-waste sites.

Within each objective, generative justice and co-design practices were implemented to ensure the partners in Ghana gain equally as much from the project as our team at WPI. Numerous tests were conducted in this project to better understand the workings of the Stirling engine. These tests were used to test the wheel speed, maximum voltage for charging cell phones, and engine efficiency. We ran additional tests on e-waste components to conclude if they were suitable replacements on the engine. Through data collection, the team demonstrated that a Stirling engine can be used to charge a cell phone. A prototype servomotor was selected from an e-waste site and incorporated into the Stirling engine assembly. The Stirling engine powered the servomotor, and the resulting voltage was used to charge a cell phone. We modeled the servomotor to evaluate performance and established performance regions for charging a cell phone quickly. Major components of the Stirling engine were designed using standard engineering computer-aided design (CAD) tools. The testing of functions and performance of the Stirling engine-powered servomotor and the charging of cell phone provided a better understanding of how the components in the engine assembly can be substituted with e-waste materials. Results from the testing and modeling guided the adjustments that can be made to maximize the voltage output for charging cell phones. Through research, experimental work, and analysis, we were able to create a prototype of the ideal Stirling engine and provided the schematics to the team of students and professors in Accra, Ghana. Further research will be continued by the Ghanian partners.

Stirling engines have been applied as energy sources since the 1870s when they were used for industrial projects in Great Britain. In the early twentieth century, they fell out of favor and have almost been forgotten about over time. Stirling engines are simple designs and they run on any source of heat and can produce a small amount of electricity. Such engines could provide a great deal of opportunity for people across sub-Saharan Africa. The application of the Stirling engine to charge cell phones in remote places of Ghana could open other opportunities including the powering of a valve in a pump for an irrigation system or running a ceiling or desk fan.

Our team collaborated with the students and professors at ACUC as well as people associated with the local e-waste management site in Accra to design and build a charging system for cell phone use in Ghana. Based on the information we provided through these conversations, the team made informed decisions about what e-waste materials could be reused and recycled to replace parts of the Stirling engine prototype. Once these decisions were made, the team worked with a local e-waste site in Massachusetts to receive e-waste materials similar to the ones available in Ghana. Additionally, our contacts at ACUC connected us with individuals who work at the e-waste sites to ensure we were properly utilizing the materials. The team then purchased an ideal Stirling engine to learn how the engine functions without the use of recycled materials. The team provided the Ghana students with drawings of the ideal Stirling engine while they awaited the shipment of their own engine. The team conducted varying engineering tests on the engine, including the operation of the engine in three regions: overdamped, underdamped, and critically damped. From these tests, the team could only charge a cell phone for a second due to resistance issues between the engine and the phone. The team concluded that the Stirling engine design needed more modifications than the team was able to make in our allotted time for the project before e-waste material substitutions could be applied. The Ghana student team took the results from our project and their access to their own ideal Stirling engine to continue their own tests and make the recommended modifications to the engine. Once the Stirling engines are designed and tested, the drawings for the engine will be distributed to individuals at the e-waste site for them to produce and sell to Accra citizens for home use.

This chapter describes the project and the relevant information researched to complete our methodology. The sections are as follows:

– Foundation of the project: the circumstances that prompted the project and what we hope to accomplish.
– Ghana's electricity: the situation in sub-Saharan Africa, specifically Ghana.
– E-waste in Ghana: the e-waste available for use and how it relates to our project.
– History of the Stirling engine: the invention as well as its prior uses.
– Stirling engine design: the common designs and the major components of the engine.
– Theoretical Stirling engine function: the process of the ideal Stirling cycle is described.
– Solar-powered Stirling engines: cases where solar energy was used to power a Stirling engine.
– Required component materials: the important aspects to consider when selecting materials for the components.

5.2 Foundation of the project and Ghana's energy challenges

The major goal of this project is to design a Stirling engine for manufacture using locally available parts in Ghana capable of charging cell phones. This design will help the majority of Ghana citizens who struggle with unreliable electricity sources. Although electricity in Ghana is widely available, it is unreliable due to chronic unpredictable power outages [3]. A combination of the Stirling engine design and the accessible Ghana e-waste site can provide the Ghanian citizens with reliable, sustainable energy. The Stirling engine was chosen as one possible solution because it is inexpensive and its design can be replicated using numerous component materials. Several of the required components can be replaced with parts from recycled appliances and electronics. This would repurpose these materials and could put them toward a sustainable and reusable application. E-waste sites have access to many secondhand parts that can be repurposed for use in a Stirling engine. Africa, including Accra, Ghana, has a large amount of e-waste materials imported from western locations. Due to Ghana's abundance of e-waste materials and their lack of reliable electricity, Stirling engines made of recycled e-waste materials could be a sustainable and accessible solution.

To make the Stirling engine an effective solution, we followed the criteria in Figure 5.2.

The project criteria enabled us to review several applications of Stirling engine. We also used the criteria to select and utilize standard components for the Stirling engine design, servomotor, and other components in the Stirling engine assembly. This criteria ignited greater attention on the cost for manufacturing the Stirling engine, cell phone charging system, and obtaining e-waste materials in Ghana.

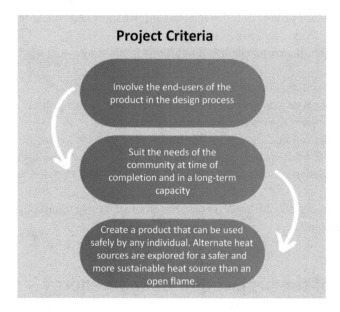

Figure 5.2: Established project criteria for realization.

5.2.1 Ghana's electricity and energy-harvesting challenges

Ghana and its people continue to try to harvest electricity in a sustained way. The growth of innovative technologies, the collaboration with the developed world, and drastic changes in the living standards greatly influence Ghana's electricity harvesting. Although there are lifestyles that continue without it, much of the world's population has shifted to using electricity for the majority of daily tasks. In wealthier or countries that the United Nations defines as developed, electricity is more widely used. This is due to easier access and a cultural dependence on it that has grown since electricity first became a part of the world. Accessibility and availability of electricity are key components in the economic well-being of individuals and countries. In sub-Saharan Africa, though, less than half the population has access to electricity as of 2018. Expanding accessibility is an important step to continued resiliency; however, this also comes at great cost. With the current rate of expansion in sub-Saharan Africa, 60% of the population will have electricity access by 2030 [4]. These numbers are significantly different from electricity access rates across the globe. In fact, 95% of people who do not have access to electricity live in sub-Saharan Africa, Latin America, and Asia [5]. These regions face similar challenges in development; however, they also have the disadvantage of no electricity access. Although increasing access to electricity may be a slow and costly process, alternative individual-use electricity sources can be used as a supplement.

In regions where electricity is accessible, many communities continue to struggle with reliable electricity. Many issues arise from unreliable electricity supply such as slowdown in industrial activity, job and income losses, and disruptions in social life [6]. In regions of poor electricity reliability, the rate of development suffers the consequences of these power disruptions. There has been an increase in consumption and demand for electricity. This, along with limited investment in generation facilities, shortfalls in electricity imports, and institutional challenges have all played a part in Ghana's electricity supply shortages and unreliability [7]. Additionally, the cost of electricity in Ghana is a problem; when compared to the average annual income, it is not affordable. Specifically in Accra, the average annual income is $883 with an average income range of $165–$1,100 [8]. According to Enerdata [9], Ghana uses 485 kWh/capita of electricity per year. This amounts to an individual's yearly total of $77.60, based on electricity price data from SmartSolar Ghana. An individual making an average income uses 11% of their money on electricity. A study shows that households can pay only 8% of their incomes for electricity [10]. Therefore, the price of electricity is not affordable to the average Ghanaian population.

5.2.2 E-waste in Ghana and business opportunities

E-waste in Ghana and its reuse to make products create new business opportunities for many Ghanians. In Ghana, the growth in demand for consumer electronics has been tied to an increase in population and the general changing of consumer habits [11]. With an increase of electronics being purchased and used globally, there comes an increase in electronic waste. For many countries, the amount of waste they are producing is too much for them to manage independently. Such countries and other regions or countries around the world that are willing to take in their e-waste in return for various goods or services [12]. Through this, a massive e-waste site in Ghana has become internationally known for being the largest e-waste site on the planet [13]. Old Fadama, pictured in Figure 5.3, is the digital waste dump in Accra, Ghana, but is more commonly known as Agbogbloshie.

It is certain that the people and communities living near this e-waste site are impacted adversely in terms of their health and the environment by the toxic substances from some of the components of electronic waste or e-waste. As consumers, businesses, and industries around the world focus on increasing their use of electronics, machines, and devices, Ghana and other developing countries will continue to receive massive e-waste for recycling purpose, creating new business opportunities and bring to end-of-life circulation.

The area where the site is in Agbogbloshie has grown into an informal settlement with a population of 8,305 people as of the 2010 census [15, p. 866]. Many individuals depend on working at the site, with it being the sole place of residence for some. Every day, the site collects trucks of e-waste from many countries around the world. It

Figure 5.3: View of Agbogbloshie, Accra, Ghana [14].

has become the job of the site workers to sift through and organize the component materials provided. The e-waste is then recycled and/or resold within Ghana or exported to other countries. The site workers create business from the e-waste and make a living from what other countries consider trash.

There are numerous risks involved with the work, disorganized system, and environmental security challenges at Agbogbloshie. It is common to hear about unregulated e-waste disposal processes. In the United States, for example, there are no national laws for managing e-waste; it is the responsibility of the state governments to implement and enforce regulations [16]. It is important to have regulations in place as the work involved in collecting and disposing of the electronic waste can be harmful to the environment and to those involved in the process. The informal economic activity in Ghana has become organized over time; associations, boards, and funding have all become implemented in the system. The Ghanaian Ministry of Environment, Science, Technology, and Innovation uses advisory services and international exchange to develop and effectively implement suitable e-waste regulations [11]. The Ghanaian government also recognizes a recently formed association of formal recycling enterprises to further this. The site in Agbogbloshie still struggles with environmental and health issues; however, so much economic empowerment and dependency lies at the hands of the workers who would not make a living without the site [17]. This abundance of e-waste present in Ghana provides an excellent source of materials that can be repurposed. This and also the inconsistency in access to electricity are two parts of the foundation of our project. The remaining aspect is the Stirling engine that will be explained in the following sections starting with its history.

5.2.3 History and applications of the Stirling engine

The Stirling engine was originally invented to compete with the steam engine in the early 1800s. Stirling engines are quiet, nonpolluting, and a reliable method to generate power. These engines use external heat sources to vary the temperature of a gas [18]. Robert Stirling, known as the father of the Stirling engine, led for the Stirling engine patent in 1816 in Edinburgh, Scotland. The patent application was titled "Improvements for diminishing the consumption of fuel and in particular an engine capable of being applied to the moving (of) machinery on a principle entirely new" [19]. Included in the patent was an explanation of the construction and use of a regenerator as well as the first-ever description of a closed-cycle, hot-air engine system [19]. What is known as a regenerator today was an economizer or heat exchanger that Stirling used in his design [18]. Figure 5.4 depicts a portrait of Robert Stirling who wanted to create a device to replace the steam engine as these engines often exploded and caused many injuries to workers [18]. Robert Stirling's engine design would be a safer alternative since it functions at lower pressures and stops if the heater section fails [18]. Another notable benefit is the engine that runs on external combustion, which can be designed to perform with high efficiency, and therefore contributes little to no air pollution [21]. Figure 5.5 shows a model of the Stirling engine from which it can be seen that the components then are much heavier than the components used in modern Stirling engine applications. James Stirling, Robert Stirling's brother and partner, presented at the Institution of Civil Engineers in 1845 and emphasized that the Stirling engine was not only aimed at saving fuel but also could serve as a

Figure 5.4: Robert Stirling [20].

safer alternative to the steam engine. In the widely used steam engines, the boilers would often explode, causing numerous injuries and deaths. Although this is not possible with the use of Stirling engines, they do not perform well at higher temperatures and fell out of commercial use due to frequent cylinder failures [19].

Figure 5.5: Model of Stirling engine at the National Museum of Scotland [20].

Another issue of the Stirling engine is the comparative size; to see similar results as a steam engine, the Stirling engine would need to be significantly larger. The cost of the production and size barriers was what eventually led to the Stirling engine becoming less successful. Because of their lack of presence at the commercial level, Stirling engines were overlooked for many years. In 1938, the Philips company took interest in the design and licensed numerous patents. In 1952, Philips had developed a power generator with the Stirling engine system as a drive. This never went into mass production, but the research work conducted by Philips later led to development of cryogenic cooling systems and other successes [19]. In recent years, Stirling engines have been used for submarines where they are able to extend the period spent underwater from a few days to weeks. The use of Stirling engines also allows for the submarine to be much quieter during travel, a large advantage for situations where concealment is necessary (nms.ac.uk, n.d.). In every design of a Stirling engine, there are a number of components essential to its function. These components as well as the necessary conditions are discussed in the following section.

5.2.4 Design and manufacture of Stirling engine components

Stirling engines use heat to produce power and are a type of external combustion engine. The average Stirling engine has seven main components: the piston, cylinder, flywheel, connecting rod, crankshaft, gas, and external heat source. These components can be arranged to form the three main types of Stirling engines: the alpha engine, the beta engine, and the gamma engine [22]. The first Stirling engine type is the alpha, which incorporates a configuration of two pistons in two different cylinders. One cylinder is connected to a cooler and the other to a heater, with the heated cylinder involved with expansion and the cooler side involved with compression as shown in Figure 5.6.

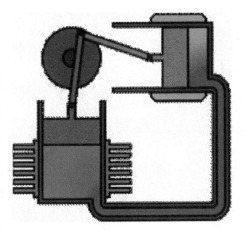

Figure 5.6: Alpha Stirling engine design [23].

As shown in Figure 5.6, the cylinders are connected with a regenerator shaft. This configuration is the simplest design of a Stirling engine, and the separation of the two cylinders prevents premature mixing of the hot and cold fluids [24]. The second Stirling engine type is the beta, which includes a singular cylinder, a power piston, and a displacer piston. The displacer piston moves the air between the cold and hot cylinders of the engine. This causes a change in the pressure between the cylinders that moves the piston back and forth as shown in Figure 5.7.

In this Stirling engine, the power piston and displacer move so that the gas compresses while passing over the cooler section of the engine and expands while it passes over the heater area as shown in Figure 5.7 [23]. The last Stirling engine type is the gamma, which has a power piston and a displacer, but unlike the beta, the power piston and the displacer are in different cylinders as shown in Figure 5.8.

The cylinder with the displacer piston also has the cooler and heater attached to it [23]. One side of the cylinder is heated, and the other side of the displacer is cooled by the cooling systems. The downside to this design is that some of the expansion process

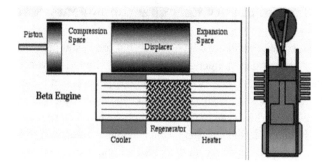

Figure 5.7: Beta Stirling engine design [23].

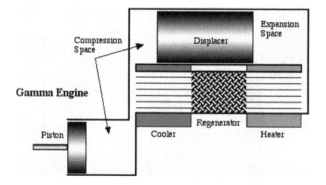

Figure 5.8: Gamma Stirling engine design [23].

has to occur in the compression space, leading to reduced output of specific power [23]. Although there are different mechanical designs of the Stirling engine, material properties for each have the same requirements. For example, the cylinders must tolerate the temperatures of the heat source without deformation as well as maintain the temperature difference between the hot and cold cylinders. The materials used in the Stirling engine are chosen to allow the engine to have its peak performance.

5.3.4 Component materials and their selection

The Stirling engine block is typically made of cast ductile iron or aluminum alloy. For low-temperature differential (LTD) Stirling engines, the internal components of the engine tend to be made of cast ductile iron or aluminum alloy; however, this can change depending on the required strength or heat resistance. As previously stated, the seven main components of a Stirling engine are the piston, cylinder, flywheel, connecting rod, crankshaft, gas, and external heat source. The piston and connecting rod are typically made of aluminum alloy due to its lightweight [25]. The power cylinders are made of brass. Another component, known as the displacer piston, can have different material that depends on the temperature differential. For example, in 1992, a Stirling engine that had a temperature difference of 1.8 °C was built, and the displacer was made of Styrofoam™. In the same engine, the piston was made of graphite [25]. In a more recent experiment, medium-density fiberboard was the highest performing material for an LTD Stirling engine as it provided more power compared to an aluminum alloy displacer [25]. The application of the engine will determine the temperature differential and thus the material needed on the engine. The temperature differential is caused by the heating and cooling of the cylinders. This process is known as a thermodynamic cycle and is explained in the following sections.

5.3.5 Theoretical Stirling engine functions and cycles

With the heating and cooling of the cylinders, the piston moves and causes the flywheel to rotate. Although all engines are designed to move a piston, Stirling engines are unique due to functioning using a closed system of working fluid [26]. Since the working fluid is in a closed system, the Stirling engine can function on an external heat source. This process is known as the Stirling cycle and is one of many thermodynamic cycles. There are several thermodynamic cycles used to transform heat energy into mechanical energy. In such heat engines, the working fluid, typically a gas, is heated, which causes the expansion that moves the piston. The major properties of thermodynamic cycles are pressure, temperature, volume, and the working fluid. From these, the work done by the engine and the efficiency of the engine can be determined. The Carnot cycle is the most efficient in Thermodynamics [27]. This is a theoretical cycle designed for the steps in the process to be adiabatic and reversible, which means throughout the cycle there is no heat or mass lost to the surroundings. Although the Carnot cycle is the most efficient, it is idealized and therefore not fully representable in a functioning engine. Although this is true of all heat cycles and engines, the Stirling cycle approaches the theoretical efficiency of the Carnot cycle and is appealing due to the ability to utilize waste heat.

The Stirling cycle consists of two ideally isochoric (constant volume) steps and two ideally isothermal (constant temperature) steps [28]. The pressure–volume (P–V) and temperature–entropy (T–s) diagram of the cycle is shown in Figure 5.9. In these figures, the four properties correspond to the mechanical function of the engine with every change in volume coinciding with the movement of the piston. The movement of the piston as it relates to the Stirling cycle is shown in Figure 5.10.

These diagrams show the thermodynamic characteristics of the Stirling engine when it is subjected to an external heat source. These characteristics are described as follows:

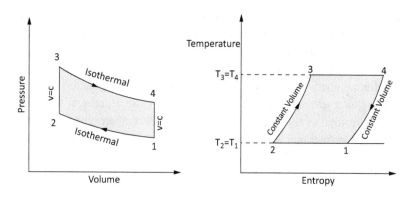

Figure 5.9: *P–V* and *T–s* diagrams of the Stirling cycle [29].

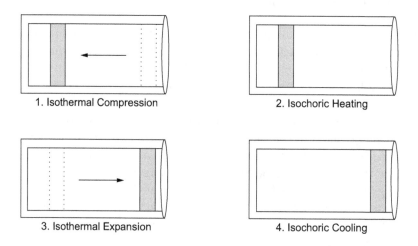

Figure 5.10: Location of the piston in the cylinder.

1. In the first step of the cycle, the working fluid undergoes isothermal compression. During this step, the pressure increases as the volume decreases and correlates to the piston moving toward the closed end of the cylinder.
2. The second step is isochoric meaning the working fluid is then kept at a constant volume as the pressure and the temperature are increased.
3. The third step is the system's second isothermal process, now with the working fluid undergoing a decrease in pressure and an increase in volume. As the gas expands, it applies pressure to the piston, which is then pushed to the end of the cylinder.
4. The last step in the cycle is again isochoric as the working fluid cools. The sliding movement of the piston can then be used to turn a flywheel and furthermore to generate a charge.

These four steps occur rapidly, providing the flywheel with thousands of revolutions per minute. Although the most used fuel in engines is a liquid combustible, Stirling engines require only a temperature difference; heat sources such as steam or the sun can be utilized by a Stirling engine [30].

Solar energy is not only sustainable but also one of the most cost-effective sources of heat. Using solar energy to power, a Stirling engine would provide a cheap and sustainable source of power. The use of solar-powered Stirling engines has existed since 1864, when John Ericsson invented a solar-powered, hot-air engine that utilized a reflector to heat the hot cylinder end of the displacer [30]. Creating a focused point of solar energy onto the hot cylinder of an alpha design-type Stirling engine provides the needed heat source for the engine to produce power. In this process, solar power is converted into mechanical energy, which can then further be converted into electrical power, thus using the solar energy to create an electrical output. Solar collectors are a special kind of heat exchanger, but the concave design must incorporate a way to trace the sun due to its angle. However, flat-plate collectors do not need to trace the sun and could be a possible application for the purposes of our project [31]. Apart from being sustainable, the use of solar power for the Stirling engine design allows for a safer heat source that is easily controlled.

5.4 Stirling engine and cell phone charger

5.4.1 Introduction

This section introduces the methods employed to achieve the goal of our project and the materials involved in those processes. We begin by discussing the objectives set by the team that aided in fulfilling the project goal. From there, we elaborate on our collaboration with the ACUC students and their contributions to this project. We then evaluate our ethical responsibilities and how they were addressed throughout the project work and report writing. With the project goal to design a Stirling engine utilizing e-waste components, we set objectives to guide the project work:

- Stirling engine investigation and reverse engineer
- Proof of concept
- Identifying e-waste materials
- Component substitution

Before any redesigning could occur, the initial Stirling engine research was done to form an understanding of the function and intent of the components in the Stirling engine. After reverse engineering an ideal Stirling engine, the team conducted tests and calculations to ensure that charging a phone was possible. Then, the available e-waste materials were investigated and assessed with respect to the project. Various components of the Stirling engine were reviewed for the replacement and substitution with recycled materials. Throughout the pursuit of these objectives, there was an open communication with the student team at ACUC to effectively utilize the co-design process.

5.4.2 Stirling engine investigation and reverse engineering

Gaining a complete understanding of the function and components of Stirling engines is necessary to make e-waste substitutions. To accomplish this, a Stirling engine kit was purchased, as shown in Figure 5.11, from Amazon.com and assembled. It provided an opportunity for our team to gain a physical understanding of the alpha Stirling engine design while also allowing for spending more time on the proof of concept and material substitution portions of the project. The alpha design of the engine was chosen due to the background research the team conducted. It was decided that the alpha design was the simplest of the Stirling engine designs. This engine will be referred to as the original engine. The original engine was considered the ideal Stirling engine, meaning it was manufactured and did not include any replacement of e-waste parts. It was chosen due to its size and advertised voltage output. This Stirling engine was stated to have a voltage output range of 5–9 V, aligning well with the 5 V minimum needed to charge a cell phone.

Figure 5.11: Assembled Stirling engine kit (original engine).

Upon receiving the original engine, the first step was to observe how it performed. This was done by finding a fuel source; in this case, 90% isopropyl alcohol. The fuel was put in the glass jar provided and the wick was soaked in the fuel. The wick was then placed under the hot cylinder and lit. After waiting an appropriate time for the cylinder to heat up the flywheel, it was manually spun to start the engine. A voltmeter was used to test the output voltage from the motor. After observing the Stirling engine's function, it was disassembled to be reverse engineered. The critical dimensions of all the components were measured using calipers and recorded on preliminary hand drawings. Using the CAD program SolidWorks, we modeled the Stirling engine.

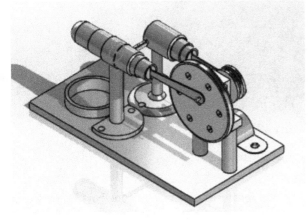

Figure 5.12: SolidWorks assembly of the Stirling engine kit.

This allowed us to communicate more effectively with the student team at ACUC in Ghana and perform finite element analysis on each component as well as the full assembly. The finished assembly is shown in Figure 5.12; additional SolidWorks models and drawings can be found in Appendix A. The CAD model shown in Figure 5.12 and those extended models of the Stirling engine in Appendix A were created using WPI engineering computer tools. Modifications were made to the original Stirling engine to ensure it can reach 5 V. This is covered in more detail in Section 5.4. After the disassembly process of the original Stirling engine, the team purchased a second engine; hereafter referred to as the new engine, from Enginediy.com, for additional testing. We utilized the same tests for the new engine that were applied to the original engine. The new engine is pictured in Figure 5.13.

This Stirling engine in this figure was used in the experiment to verify the production of the required voltage by the Stirling engine for charging cell phones. The data collected through testing was used in the component substitution process to ensure the properties of the new materials were comparable to the replaced components.

5.4.3 Proof of concept and co-design activities

Proof of concept meant running tests and calculating if the ideal Stirling engine we purchased could charge a cell phone to achieve our goal. These tests included measuring the surrounding air temperature and running the engine to see if it had higher performance

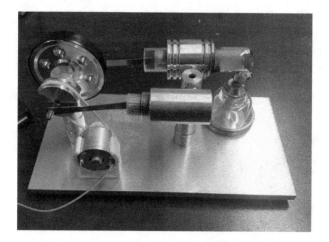

Figure 5.13: New Stirling engine kit purchased (second engine).

in colder or warmer temperatures. We conducted tests to determine how long it would take for the engine to get to maximum voltage, and then after extinguishing the flame, how long it took for the voltage to reach 0 V. Lastly, we performed several tests to experimentally ascertain the angular velocity of the flywheel. This expands our understanding of the relationship between the thermodynamic cycle and the function of the ideal Stirling engine.

To test the efficiency of the Stirling engine in varying temperatures, the team used a thermometer to test the air temperature while running the engine. These tests were run in varying environments, including a heat-controlled room and the outside temperature on days of varying weather. When testing how long it took the engine to reach its maximum output voltage, the team started by using a stopwatch application on our mobile devices and clicked start once the fire was lit underneath the hot cylinder. We then proceeded to spin the flywheel every 5 s to test how long it took for the hot cylinder to heat up enough for the engine to start moving on its own. Once the engine began to move autonomously, a lap was added on the stopwatch. The team then observed the voltage output until it reached a consistent maximum voltage. The stopwatch was lapped again and the time it took to reach its maximum was recorded. The fire was then extinguished and the stopwatch continued to run until the voltage output was zero and the flywheels were stationary. All results were recorded.

The last test we conducted was a flywheel speed test. The team performed the flywheel speed test simultaneously with the time tests even though they are independent from one another. The speed test was done by using a tachometer device. The tachometer works by tracking a reflective piece of tape as it produces a laser aimed at the target of the readings. Due to the flywheel itself being reflective, the team used black electrical tape to cover the side of the flywheel to be measured to ensure that the reflective material would not disrupt the readings.

The team placed a piece of reflective tape over the electrical tape and started the engine, as shown in Figure 5.14. This figure shows another view of the Stirling engine that was used in the laboratory for testing. The measurements were taken once the engine reached its maximum output voltage by aiming the tachometer over the flywheel where the reflective tape was attached. All results were recorded. The Stirling cycle is a thermodynamic process used on Stirling engines and other Stirling devices.

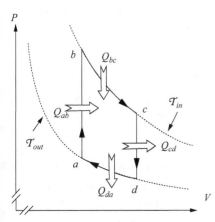

Figure 5.14: The Stirling engine prepared for flywheel speed testing.

Figure 5.15: The Stirling cycle's *p–v* diagram.

Figure 5.15 illustrates these cycles. In this figure, the parameter p is the pressure and v is the volume. Also, this figure is referred to as pressure–volume or p–v diagram for the Stirling cycle. As previously stated, this cycle contains isochoric and isothermal processes. In an isochoric process, the volume stays constant and the temperature and pressure change. In an isothermal process, the temperature stays constant and the volume and pressure change. During the isochoric process, since the change in volume is zero, there is no work done W, given by the equation

$$W = \int_{v_i}^{v_f} pdv, \tag{5.1}$$

where p is the pressure, v_i is the initial volume, and v_f is the final volume. From the gas equation $pv = nRT$, we express the volume v as a subject of the formula and substitute the result into the work done, equation (5.1) leads to

$$
\begin{aligned}
W &= \int_{v_i}^{v_f} pdv \\
&= nRT \int_{v_i}^{v_f} \frac{1}{v} dv \\
&= nRT \left(\ln v_f - \ln v_i \right) + \ln C \\
&= nRT \ln \left(C \frac{v_f}{v_i} \right).
\end{aligned}
\tag{5.2}
$$

In this equation, $\ln C$ is the constant of integration, n is the amount of working fluid in moles, R is the ideal gas constant, and T is the temperature. Given the work formula in equations (5.2), it can be assumed that all the work done in the Stirling cycle is completed during the isothermal process. Solving for the constant of integration for zero boundary condition yields the equation for the mechanical work:

$$W = nRT \ln \left(\frac{v_f}{v_i} \right). \tag{5.3}$$

Within the natural log is the compression ratio, the ratio of the maximum volume of the engine to the minimum. This simplifies the calculation of the engine's thermal efficiency, which is given by the equation $= W/Q$, where Q is equal to the heat input to the engine to allow for continuous cycling. The International System (SI) unit for work, energy, and heat is joule which is kilogram-meter squared per second squared or Newton meter. The power p that the engine produces can be found using the equation $p = !$, where the parameter $!$ is the angular speed of the engine's flywheel and is the torque. The magnitude of the torque is found by multiplying the tangential force F by the perpendicular distance d to the force. The SI unit for power is Newton meter per second, which is watt, and it is named after James Watt who exuded enormous effort to improve the performance of steam engines. The horsepower is commonly used as a unit for power and 1 horsepower is equivalent in the USC (US Customary) system of units as 3.3×10^4 ft.lb/min or $5.50 \times 10^2 \frac{ft.lb}{sec}$. The heat input Q into the Stirling engine is the sum of all the heat quantities > zero. This can be found using the equation $Q = mc\ T$, where m is the mass of the working fluid; in this case, c is the isovolumetric heat capacity of air, and T is the change in temperature of the Stirling engine system. Since the full output of the heat source will be unevenly received by the Stirling engine, this equation is used to approximate the total heat input.

Generally, physical quantities may be measured in both the *SI and USC or British System of Units.* Changing from the SI to the USC system of units is achieved by the conversion formula *SI=USC.* An inch is 2.54 cm, 1 ft = 0.348 m, and 1 km = 0.592 mile. The conversion of feet to inches is 12 in = 1 ft, meter to kilometer is 1,000 m = 1 km and hour to seconds is 1 h = 3,600 s. The ratio SI = USC = is unique since it has the same dimensions in the SI and UCS systems of units. A dimension is the basic measure of a physical quantity, and it has a number with a unit. One meter of length in the SI system is equal to 3.281 ft in the USC unit system and 39.37 in is 1 m. A kilogram is the SI unit for mass, and it is expressed in the USC unit system as a slug using the conversation ratio slug = kg = 14.59. A slug is $lb.s^2$ = ft and one slug is 1.46×10^4 g. The USC unit of weight is pound when the gravitational acceleration $g = 32.2$ ft = s^2 and mass in slug. Consistency of units in equations and constitutive relations is essential for obtaining rational answers to a problem. It is suggested to always verify the units of quantities in the equations and constitutive relations before numerical values are substituted for the parameters.

5.4.4 Identifying and evaluating e-waste materials

Before the Stirling engine components can be substituted with e-waste, an understanding of the materials available had to be achieved. To accomplish this, we visited a local e-waste site, spoke with an e-waste worker in Ghana, and gathered materials for assessment. The visit to the Wachusett Watershed Regional Recycling Center in West Boylston, Massachusetts, provided us with first-hand knowledge of the available materials and their condition. We assumed that the e-waste available to us locally is similar to what is nearby to our partners in Ghana. To confirm this assumption, we asked Julian Bennett, the advisor to our team from ACUC, to help us contact Suali, an e-waste site worker in Accra. The conversation with Suali confirmed our assumptions and informed us of the many other available materials – some as large as cars. Armed with this knowledge, we revisited the Wachusett Watershed to collect materials. We evaluated the state of the gathered items, material composition, working condition, and ease of disassembly. These assessments were used in conjunction with the knowledge gained in the prior sections to evaluate the e-waste materials with the intent of component substitution. With the knowledge of how each component functions, the required properties for each component, the available e-waste and its condition, and suggestions for part replacement were made. Using materials from the e-waste site, individual components were substituted into the Stirling engine kit and tested for performance. With each part we collected from the e-waste site in Massachusetts, we discussed them with the Ghana team to gain a broader perspective for the redesign process. We asked what they thought of our ideas and asked for additional suggestions for the substitution process. We then began testing parts once we became more confident about how it could fit into the redesign. The tests included temporarily replacing a component with a recycled part on the engine somewhere and running the engine. We then compared the results from that test with assessments where the recycled component was not present to see if it made a difference. Some recycled parts we found and tested included direct current (DC) motors from a CD player and a toy car, rubber bands of various sizes and thicknesses, various thick metals from computers and other devices, and heat sinks from computer circuit boards.

5.4.5 Co-design with Academic City University College students

Co-design was highly integrated into the project and was the main focus in the design and outcome of the process. The Ghana team was consulted, and they helped throughout almost the entire project. They made suggestions to our material selections for substituting parts, they assisted in modeling the CAD and received the same engine so that they could be involved in the testing. This process was important for both teams to understand how all the parts fit together as well as understanding what types of e-waste materials could be used to replace them. Co-design was important for this project because the outcome and final product affect the citizens of Ghana, which is why involving them in the process was essential. Having a Ghana team as partners helped bring the background and knowledge of the community, culture, and life experiences that differed from our own. The team also partnered with workers at Agbogbloshie, which helped them gain a deeper insight into the site. It also provided both teams with valuable information about what materials were available for the Ghana team to use when reconstructing and remodeling the engine out of the e-waste materials. These materials are also accessible to most people living in Accra. Many structural substitutions can be left up to the Ghanaian people who will be recreating the engines for their own personal use. This allows them to incorporate their own parts and be involved in the making of the engines they will use. This permits the design of the engine to be malleable and flexible so that there are many outcomes and possibilities for its design. This helps with the co-design process and allows for the project to be sustainable and long-lasting even after the WPI team completes their portion of the work on the project.

5.4.6 Ethical engineering and method of data collection

Ethical engineering is important as the work of engineers has copious impacts on the quality of life for all people. The role of conducting ethical work is to make it easy for people to trust and value your project. In engineering, there are ethical considerations to take when conducting research and testing. The Code of Ethics for Engineers is a professional guide that all engineers hold each other to follow. The guide outlines standards, duties, and practices to ensure that the services engineers provide are honest, impartial, fair, equitable, and safe for all [32]. For the same reason that professional engineers have a code of ethics to follow, and universities have a student code of conduct to follow, our team wanted to ensure we safely conducted our work. Although we did not have a written set of guidelines to follow, we still held each other accountable for our actions and practiced safe methods. As we are working with others outside our group, we respect their choices and rights not to participate if they choose. We asked for consent to use their names in the paper, which would be published, and in the case of wanting to stay anonymous we respect the rules of confidentiality. These partners are supporting our efforts to complete this project and, therefore, we support their decisions. This focus on ethical work allows us to work together and make sure all parties are comfortable. All the individuals we partnered with in this project are given credit for their contributions, whether that be their knowledge, time, finances, or guidance. Without the support from our partners, this project would look very different and we must acknowledge those contributions, or it would be unfair. It is important to ethically conduct projects to ensure the data and information you are collecting is not wrongfully gathered as you want your project to be valued and respected.

5.4.7 Stirling engine-powered servomotor and modeling

An important objective for the project was using the Stirling engine to charge cell phones. A select number of e-waste servomotors were tested and evaluated. The Stirling engine provided the input voltage to the series servomotor to produce an output current capable of charging a cell phone for limited amount of use. Alternating current (AC), DC, brushless DC, gear DC, and stepper motors are used to power several engines, machines, and devices. Power companies supply AC, and in the United States, the AC is distributed by alternating sinusoidal voltage and current at a frequency of about 60 Hz and peak voltage of about 120, 240, or 480 V. An AC motor is composed of several nonlinear elements, transitions, and variations that make the application of AC motors more complex than a DC motor. A DC motor converts electrical energy into mechanical energy, while a DC generator converts mechanical energy into electrical energy. A DC motor is composed of a power voltage amplifier, electric circuit, magnetic field coil, brushes, and mechanical load. The amplifier supplies voltage to the electric circuit and that voltage supply is used to generate the field current flowing through the coil of wire. Brushes are attached on the rotating commutator, and they act as resistors against the current-carrying conductors. Brushes are also noise absorbers, and they cause the DC motor to run more quietly. The torque developed by the motor is proportional to the field current and this torque drives the motor shaft. A DC motor can be (a) series configuration, (b) parallel configuration, and (c) combined series and parallel configuration. Each configuration provides different rated torque and speed characteristics. A series configuration is referred to as a series DC motor. This type of DC motor is known to produce torques for a range of speed and load variations as the flow of current increases. A series DC motor is found in many applications, including robotics, medical devices, and mining machines. A parallel configuration DC motor is called a shunt motor. These motors have two or more paths for the current to flow through the circuit elements. The torque produced by the current of one path is independent of the current of the other path. Shunt motors are found in air conditioners, snow blowers, and temperature control fans. The combined series and parallel configurations are called a compound DC motor. In a compound DC motor, the series winding will increase the strength of the magnetic field for the shunt winding to respond to increasing loads. The amount of speed variation in a compound DC motor depends on the magnetic field strength of the series DC motor. The brushes on the rotating commutator of a DC motor act as resistors against the current-carrying conductors and they are also noise absorbers for the DC motor to run more quietly. Gear motors composed of gear boxes are available for reducing speed and torque variations as current flows through the circuit elements. Flywheels, valves, springs, dampers, pulleys, linkages, and crankshafts are also available to smooth motor torque and speed variations. The mechanical energy of the Stirling engine from the thermoheat source is stored into the flywheel, and this mechanical energy is refined and transferred to the servomotor through the piston–crankshaft and the driving load.

5.4.8 Formulating servomotor equations and analysis

DC motors of series were widely used a variety of applications. These kinds of DC motors are typically equipped with stabilizing circuit devices to reduce speed–torque variations as the effective moment of inertia of the rotating components on the output motor shaft is varied. The incorporation of a series servomotor into the Stirling engine assembly converts a thermo-heat supply to a current–voltage power supply for charging cell phones in Ghana. For the series servomotor, the current flowing through the circuit elements is the same, while the voltage drop across each element in the circuit is different. The input thermovoltage power by the Stirling engine produces a current to flow through a coil of wires inside the series servomotor. As the current and strength of the magnetic field in the servomotor increase, the difference between the power supply voltage and sum of all the voltage drops across the components of the electric circuit changes. Figure 5.16 shows the electric circuit of the servomotor in the Stirling engine assembly. The electric circuit is made up of an applied voltage $v_a(t)$, armature resistance R_a, armature inductance L_a,

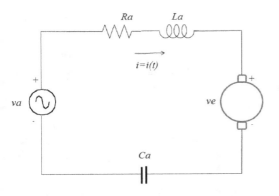

Figure 5.16: Electric circuit of the armature DC servomotor.

armature load capacitance C_a, transducer voltage $v_e(\cdot)$, and armature current $i_a(t)$. The purpose of capacitor C_a is to store and release energy at a moment in the Stirling cycle and as required by the voltage difference between the applied voltage $v_a(t)$ and the transducer voltage $v_e(\cdot)$. In the series $R_a C_a L_a$ electric circuit, the components for the servomotor in the Stirling engine are connected with the applied voltage supply $v_a(t)$ and back emf voltage or transducer voltage $v_e(\cdot)$. The transducer voltage $v_e(\cdot)$ is proportional to the angular speed $! = \frac{d\theta}{dt} = \_$ of the motor shaft. In addition to the circuit element shown in Figure 17, the motor shaft angle, angular velocity $! = \_$, equivalent torsional elastic spring k, and bearing and lubricant damper c are present. Since the applied voltage supply $v_a(t)$ by the Stirling engine, armature current $i_a(t)$, resistance R_a, inductance L_a, load capacitance C_a, and voltage transducer $v_e = v_e(\cdot)$ are connected in series, then by Kirchhoff's voltage law (KVL), the applied voltage supply $v_a(t)$ is equal to the sum of all the voltage drops across the resistance R_a, inductance L_a, capacitance C_a, and transducer voltage $v_e(\cdot)$. v_{Ra} denotes the voltage drop across R_a, v_{La} the voltage drop across L_a, and v_{Ca} the voltage drop across C_a, and KVL gives the algebraic equation

$$v_L(t) + v_R(t) + v_C(t) + v_e(-) = v_a(t) \tag{5.4}$$

for the series servomotor. Ohm's law establishes the relations for the voltage drops in equation (5.4) in terms of the armature current $i_a(t)$ of the servomotor. In the presence of the resistance R_a, inductance L_a, and capacitance C_a in the servomotor, the Stirling engine will produce a current–voltage power supply at speeds proportional to the ratio of L_a to R_a and to the product of $R_a C_a$. The capacitor in the servomotor is also turned on based upon the amount of current flow through the terminals and heat loss to the atmosphere. When the capacitor $C_a = 0$, the speed at which the Stirling engine can produce voltage required for charging a cell phone depends upon the ratio of the inductor L_a to the resistor R_a.

Current is defined as the time rate of flow of electric charge, and mathematically this relationship is written as $i_a = \frac{dq(t)}{dt} = \dot{q}$, where the parameter $q = q(t)$ denotes the electric charge in coulombs. Voltage is measured in volts and current is measured in amperes, which is coulombs per second. The servomotor torque $_m(t)$ is proportional to the armature current $i_a = i_a(t)$; and the constant of proportionality k_t is referred to as the torque or servomotor constant. With the current $i_a(t)$ and motor torque constant k_t, we have the torque–current relation $_m(t) = k_t i_a(t)$ that drives all the mass moment of inertia mounted on the servomotor shaft. The current, torque, and speed of the servomotor are also influenced by the voltage difference $v_a(t)$ $v_e(\,)$. The voltage at the capacitor is defined as $v_{Ca} = \frac{q(t)}{C_a}$ with $C_a = 06$. Across the resistance R_a, Ohm's law gives the voltage drop as $v_{Ra} = R_a i(t)$, where the unit of resistance is Ohm, which is volt per ampere. Resistance impedes the flow of current, and its mechanical engineering analog is damping. The inductance voltage according to Ohm's law is proportional to the time rate of change of current and is written as $v_{La} = L_a \frac{di_a}{dt}$, where the inductance L_a is the constant of proportionality. Volt second per ampere is the unit of the inductance and is called Henry. Separating the variables in the expression $i_a = \frac{dq(t)}{dt}$ to obtain the differential form $dq = i_a dt$, integrating both sides of this differential form, and applying the relation $v_{Ca} = \frac{q(t)}{C_a}$ to the result yield the Ohm's law for the capacitance

$$v_{Ca}(t) = \frac{1}{C_a} \int_{t_0}^{t} i(t)dt + v_{Ca}(t_0), \quad v_{Ca}(t_0) = \frac{q_0(t_0)}{C_a}, \quad C_a \neq 0, \tag{5.5}$$

where the initial electric charge $q_0(t_0)$ and initial capacitor voltage $v_c(t_0)$ at the capacitor are values at the initial time t_0. Capacitors are current carrier copper wires and have dimensions of coulomb per volt. One coulomb per volt is equivalent to 1 F. The reciprocal of a capacitor corresponds to an elastic stiffness. Inductance manipulates the rise and fall of a current in the magnetic field and it is measured in Henry. Inductance has the dimensions of volt-second per amperes, and it corresponds to the mass of a mechanical load. Voltage is the mechanical equivalent of a force, current is equivalent to velocity, and electric charge corresponds to displacement. The product of a circuit voltage and current gives the electric power and the rate of change of work is the mechanical power.

Table 5.1 presents quantities, notations, parameters, and terminologies that will appear in the model equations of the servomotor. Also depicted in the table are the effective spring constant k, undamped natural frequency $!_n$, effective damping coefficient c, critical damping coefficient c_{crit}, and damping ratio. Dimensional analysis is an important way for verifying that the units of quantities are consistent. The SI unit system is also called the MKS system where M is meter, K is kilogram, and S is second.

Centimeter, gram, and second (CGS) is another name for the SI system of units. One meter is 100 cm, and 1×10^6 mm is equal to 1 km. The dimensions of mass $[M]$, length $[L]$, and time $[T]$ are the MKS dimensions while the foot $[F]$, pound $[P]$, and second $[S]$ are the gravitational dimensions in the USC system. A physical quantity in the notation $[:]$ denotes the dimensions of that physical quantity. The FPS (foot, pound, second) expresses length in feet, weight or force in pounds, and time in seconds. Weight has dimensions of $[MLT^2]$ and the expression $m = W = g$ also gives the dimension of the mass. A dimensionless quantity is a quantity without a unit and its dimension is 1. Time is

Table 5.1: Physical quantities and their notations and units.

Quantity	Parameter	Units
Applied voltage	$v_a = v_a(t)$	Volts
Armature current	$i_a = i_a(t)$	Amperes
Electromotive voltage or transducer voltage	$v_e = v_e(-)$	Volts
Armature resistance	R_a	Ohms
Armature capacitance	C_a	Farad
Armature inductance	L_a	Henry
Series $R_aC_aL_a$ motor	Current is same	Voltage is different, KV L
Shunt $R_aC_aL_a$ motor	Current is different	Voltage is same, KCL
Angular displacement	$= (t)$	Radians
Shaft angular velocity	$! = !(t) = !\_(t)$	rad = s
Angular acceleration	$= (t) = (t)$	rad = s^2
Motor torque	$= (t) = k_t i_a(t)$	Derived units
Torque motor constant	k_t	Derived units
Moment of inertia	J	Derived units
Effective mass	m	Derived units
Effective spring constant	k	Derived units
Effective damping coefficient	c	Derived units
Critical damping coefficient	c_{crit}	Derived units
Undamped natural frequency	$!_n$	rad = s
Damping ratio		dimensionless

measured in seconds, minutes, and hours. One minute is 60 s, 60 min is 1 h, and 1 h is 3,600 s. The dimension of time in seconds, minutes, and hours is [T]. The unit of the arc length of a circle involves length and angle in radian. A radian is the unit of an angle, and its dimension is 1. A radian is defined as the ratio of the arc length to the radius of the circle making an angle with a horizontal axis. One radian is expressed in degrees as 1 rad = 3,600 = 2 = 57.3°, 1° = 180, and 1 revolution is 2 rad.

Substituting the expressions of the voltage drops $v_{La} = L_a \frac{di_a}{dt}$ for the inductance L_a, $v_{Ra} = R_a i$ for the resistance R_a, and $v_{Ca}(t)$ in equation (5.5) for the capacitance C_a into the KVL algebraic servomotor equation (5.4) yields the servomotor integral–differential equation

$$L\frac{di(t)}{dt} + Ri(t) + \frac{1}{C}\int_{t_0}^{t} i(t)dt = v_a(t) - v_e\left(\dot{\theta}\right) - v_C(t_0). \tag{5.6}$$

Using the definition $i(t) = \frac{dq(t)}{dt}$, and notations $c_a = 1 = C_a$ and $f(t) = v_a(t)\, v_e(-)\, v_C(t_0)$, we obtain the second-order ordinary differential equation (ODE)

$$L_a\frac{dq^2(t)}{dt^2} + R_a\frac{dq(t)}{dt} + \kappa_C q(t) = f(t), \quad q(t_0) = q_0, \quad \dot{q}(t_0) = v_0, \tag{5.7}$$

where $c = 1 = C$ is referred to as the elastance. The initial electrical charge $q(t_0) = q_0$ and current $q\_(t_0) = v_0$ are the values at the initial time t_0: equation (5.7) is a second-order

ODE with constant coefficients, and the highest derivative appearing in this ODE is 2. Incidentally, there will be two arbitrary constants in the general solution of the second-order ODE and the two initial conditions $q(t_0) = q_0$, and $q\_(t_0) = v_0$ are required to obtain a particular solution. The basic shape of the power supply voltage $f(t) = v_a(t) v_e(\cdot) v_c(t_0)$ of the servomotor can be a constant, variable, sinusoidal, and nonlinear functions. The selection of a particular function for the power supply voltage is based upon the voltage supplied by the Stirling engine and speed–torque requirements for driving the loads on the servomotor output shaft.

The servomotor equation (5.7) is written to a desired standard form by expressing the undamped natural frequency $!_n$ and damping ratio in terms of the armature resistance R_a, inductance L_a, and capacitance C_a. For a mechanical system involving an effective mass m, spring constant k, and damping coefficient c, the undamped natural frequency $!_n$ is defined as $!^n = \sqrt{\frac{k}{m}}$, damping ratio is $= \frac{c}{c_{crit}}$, and critical damping $c_{crit} = 2pmk = 2\,m!_n$. These relations correspond to the single degree of mass-dashpot-spring system, which is analogous to the electrical R_a, C_a, L_a circuit of the servomotor in the Stirling engine assembly. Dividing both sides of the servomotor equation by L_a, setting $!^n = \sqrt{\frac{K_C}{L_a}}, c_{crit} = 2L_a!_n, = \frac{c}{c_{crit}} = \frac{R_a}{2\omega_n L_a}$, and simplifying, transform the servomotor equation into the desired standard form

$$\frac{dq^2(t)}{dt^2} + 2\xi\omega_n \frac{dq(t)}{dt} + \omega_n^2 q(t) = \omega_n^2 r(t), \quad q(t_0) = q_0, \quad \dot{q}(t_0) = v_0, \tag{5.8}$$

where $r(t)$ is the applied voltage and it is now defined as $r(t) = \frac{f(t)}{K_C}$ with $f(t) = v_a(t) v_e(\cdot)$ $v_c(t_0)$. In the servomotor equation (5.8), the undamped natural frequency $!^2_n = !^n = \sqrt{\frac{K_C}{L_a}}$ damping ratio $= \frac{R_a}{2\omega_n L_a}$, applied voltage $r(t)$, and armature current $i_a(t) = \frac{dq(t)}{dt}$ are present. The ratio of the Laplace transform of the applied voltage $r(t)$ to the Laplace transform of the armature current $i_a(t)$ with all the initial conditions in the servomotor being zero is the impedance, denoted by $Z = Z(s)$, and it is a function of the Laplace transform variable s. The Laplace transform variable s is a complex number that is contained in the complex space C and it has real numbers belonging to the real line $R = (1,1)$. The impedance $Z(s) = \frac{\mathcal{L}(r(t))}{\mathcal{L}(i_a(t))} = \frac{R(s)}{I_a(s)}$ of the servomotor in the Stirling engine corresponds to the sum of all the individual impedances in the series circuit configuration, where $L(r(t)) = R(s)$ is the Laplace transform of the applied voltage $r(t)$ and $L(i_a(t))$ $= I_a(s)$ is the Laplace transform of the armature current $i_a(t)$. The parameter R_a denotes resistance, inductance L_a, capacitor C_a, voltage input $v_a(t)$, and current $i_a(t)$. Using the Laplace transform method, the impedance for resistance is $Z_{Ra}(s) = R_a$, inductance is $Z_{La}(s) = L_a s$, and capacitance is $Z_C(s) = 1 = C_a s$. In the next section, the definitions and formulas for Laplace and inverse Laplace transforms are presented. Tables of Laplace and inverse Laplace transforms for selected functions are also given.

5.4.8.1 Mathematical methods and theoretical analysis

We use the mathematical method of Laplace transform and inverse Laplace transform to obtain solutions for the servomotor ODE (5.8) and impedance $Z(s) = \frac{\mathcal{L}(r(t))}{\mathcal{L}(i_a(t))} = \frac{R(s)}{I_a(s)}$, with $I_a(s) = 06$. Laplace transform deals with the transformation of a piecewise continuous function $f = f(t)$, and of exponential order on the interval $(0,1)$ $2R = (1,1)$ to a new set of functions $F = F(s)$, $s > 0$ in the complex space C by means of the definition

$$F(s) = \mathcal{L}(f(t)) = \int_0^\infty e^{st}f(t)dt = \lim_{\beta \to \infty} \left(\int_0^\beta e^{st}f(t)dt \right), \quad (5.9)$$

for all values of $s > 0$ such that the limit of the proper integral exists. The notation $F(s) = L(f(t))$ denotes the Laplace transform of the original function $f(t)$; $t\,0$ and s is the frequency domain variable. The time domain variable $t\,2R$ is a real number, while $s\,2C$ is a complex number. A number written as the form $s = +j!$ is called a complex number, where $j = p\,1$ is referred to as the imaginary unit, s is the real part and $!$ is the imaginary part. The number $s = j!$ is called the complex conjugate of $s = +j!$, and the multiplication of these two complex numbers $s = +j!$ and $s = j!$ yields the result $(+j!)(j!) = {}^2 + !^2$. We think of the exponential function e^{st} in equation (5.9) as a weighting function that slows down the growth rate of $f(t)$ for $t\,0$, where the base e for st is the base for the natural logarithm of $\ln e^x = x$. To determine $F(s) = L(f(t))$, we integrate the proper integral as t varies from 0 then take the limit of the derived result as $!1$. Integration from 0 to 1 is referred to as improper integral. Capital letters are typically used to denote the Laplace transform of a function. The Laplace transform of the functions $f(t)$, $v_a(t)$, and $i_a(t)$ are $L(v(t)) = F(s)$, $L(v_a(t)) = V_a(s)$, and $L(i_a(t)) = I_a(s)$, respectively. Laplace-transformed functions in terms of the Laplace transform variable s are called frequency functions and they are expressed in the frequency domain. The functions depending upon the time t are called time functions and they are defined in the time domain. The Laplace transform is also an important method for solving differential equations with constant coefficients. This method enables one to solve differential equations together with the initial conditions without going through multiple solution steps.

The Laplace transform and inverse Laplace transform methods are used to obtain explicit expressions for the servomotor current $i_a(t)$, torque $_m(t) = k_t i_a(t)$, and voltage power supply for the charging system. Units of quantity measure are not typically assigned to Laplace-transformed expressions depending upon the Laplace transform variable. The impedance $Z(s)$ is a transfer function of a linear electric circuit, and it is defined as the ratio of the Laplace transform of the voltage to the Laplace transform of the current with all the initial conditions in the electrical circuit being zero. The inverse Laplace transform is the process of converting the frequency function $F(s)$ in terms of the Laplace transform variable s to the time function $f(t)$. This is achieved by using the formula

$$f(t) = \mathcal{L}^{1}(F(s)) = \frac{1}{2\pi j} \int_{\sigma j\infty}^{\sigma+j\infty} F(s)e^{st}ds, s \in ,\mathbb{C}\ t \in \mathbb{R}, \tag{5.10}$$

for which routine algebra cannot easily produce the inverse Laplace transform $L^1(F(s))$ of $F(s)$. In the Laplace transform formula (5.9), the notation $L(f(t))$ donates the Laplace transform of the time function $f(t)$, and in formula (5.10) for the inverse Laplace transform, the notation $L^1(F(s))$ represents the inverse Laplace transform of $F(s)$. Admittedly, using formula (5.10) to find the inverse Laplace transform of $L^1(F(s)) = f(t)$ is not a straightforward calculation. We often rely on the use of tables of Laplace and inverse Laplace transforms. Tables 5.2–5.4 list several selected functions and their corresponding Laplace and inverse Laplace transforms. Partial fraction expansion and algebraic simplifications play a major role in finding a match of frequency functions in the tables of Laplace and inverse Laplace transforms.

In Tables 5.2, constant, impulse, and polynomial functions, which are typically used to describe the shapes of the voltage and current in a servomotor, are transformed from a time domain variable to a frequency domain variable. The use of the Laplace transform tables provides a convenient way to obtain the inverse Laplace transform $f(t) = L^1(F(s))$ without the use of improper integration formula (5.10). Pairs of

Table 5.2: Constant, polynomial, and impulsive functions.

Typical Functions	Time Functions $f = f(t), t \geq 0$	Frequency Functions $F(s) = \mathcal{L}(f(t)), s > 0$
formula	$\mathcal{L}(f(t)) = \int_0^\infty e^{-st}f(t)dt,$	$F(s)$
homogenous	$cf(t), c = [c_1, c_2] \in \mathbb{R}$	$cF(s)$
n 1	$c_1f_1(t) \pm c_2f_2(t)$	1 sk
k 1	1	
	t	
	t^2	
	t^3	
	$t^n, \ n = 1, 2, 3, \ldots$	
	—	
	—	
	$\frac{1}{\Gamma(k)}t^{k-1}, \ k = 1, 2, 3, \ldots$	
	$\delta(t)$	
	$\delta(t-a)$	
	$\delta(t-1)$	
	$\delta(t-2)$	
	$\delta(t-3)$	
	$n-$	
	$\frac{1}{(n-1)!}t^{n-1}, \ n = 1, 2, 3, \ldots$	$\frac{1}{s^n}$

Table 5.2 (continued)

Typical Functions	Time Functions $f = f(t), t \geq 0$	Frequency Functions $F(s) = \mathcal{L}(f(t)), s > 0$
linearity	±	$c_1 F_1(s) \pm c_2 F_2(s)$
unit step		$\frac{1}{s}$
unit ramp		$\frac{1}{s^2} = \frac{1}{s}\mathcal{L}(1)$
parabolic		$\frac{2}{s^3} = \frac{1}{s}\mathcal{L}(t)$
cubic		$\frac{6}{s^4} = \frac{3}{s}\mathcal{L}(t^2)$
– polynomial	—	—
– polynomial	–	—
– polynomial	–	—
unit impulse	δ	1
shifting impulse	δ –	e^{-as}
shifting impulse	δ –	e^{-s}
shifting impulse	δ –	e^{2s}
shifting impulse	δ –	e^{-3s}

time functions and s-dependent functions $F(s)$ are available in the Laplace transform table. The Laplace transform of an applied unit voltage to the Stirling engine is $L(1) = F$ $(s) = 1 = s$. If the applied voltage to the Stirling engine is the unit ramp function $f(t) = t$, then its Laplace transform $L(t)$ is $F(s) = 1 = s^2$. The Laplace transform of the Δ or impulse voltage function $f(t) = (t)$ is $L((t))$ is $F(s) = 1$. The inverse Laplace transform $L^1(1)$ is the Δ or impulse function (t). When the frequency function $F(s) = s$, the inverse Laplace transform $\mathcal{L}^{-1}(s)$ is $\mathcal{L}^{-1}(s) = \frac{d\delta(t)}{dt}$, and for $F(s) = s^2$, the inverse Laplace transform is

$$\mathcal{L}^1(s^2) = \frac{d^2\delta(t)}{dt^2}.$$

In Table 5.3, the Laplace transforms of exponential, sinusoidal, and hyperbolic functions are presented. The Laplace transform expressions in both Tables 5.2 and 5.3 are obtained using the formula equation (5.9) of the Laplace transform. The algebraic properties involved in the definitions of Laplace and inverse Laplace transforms further simplify the collection of functions for the Laplace transform tables. According to the final value theorem, the Laplace transform and inverse Laplace transform are related by the expression $\lim f(t) = \lim sF(s)$ $t!1$ $s!0$ that can determine the growth and decay rates of the circuit voltage and currents of the servomotor. The initial values for the circuit elements of the servomotor can be quantified using the initial value theo-

Table 5.3: Exponential, sinusoidal, and special functions.

Typical Functions	Time Functions	Frequency Functions
	$\geq$	$F(s) = \mathcal{L}(f(t)),\ \ s > 0$
growth exponential		$\frac{1}{s-a},\ s \neq a$
decay exponential		$\frac{1}{s+a},\ s \neq -a$
multiplicative		$\frac{1}{(s\ a)^2},\ s \neq -a$
multiplicative	$-$	$\frac{1}{(s+a)^2},\ s \neq a$
multiplicative		$\frac{n!}{(s+a)^{n+1}},\ s \neq -a$
multiplicative	$-$	$\frac{1}{(s-a)^n},\ s \neq a$
gamma function	$-$	$\frac{1}{(s-a)^k},\ s \neq a$
		$\frac{s}{s^2+\omega^2},\ s^2 + \omega^2 \neq 0$
		$\frac{\omega}{s^2+\omega^2},\ s^2 + \omega^2 \neq 0$
damped cosine		$\frac{s+a}{(s+a)^2+\omega^2},\ (s+a)^2 + \omega^2 \neq 0$
damped sine	$-$	$\frac{\omega}{(s+a)^2+\omega^2},\ (s+a)^2 + \omega^2 \neq 0$
excited cosine		$\frac{s^2-\omega^2}{\left(s^2+\omega^2\right)^2+\omega^2},\ s^2 + \omega^2 \neq 0$
excited sine		$\frac{2\omega s}{\left(s^2+\omega^2\right)^2},\ s^2 + \omega^2 \neq 0$
	$e^{at},\ a > 0$	
	$e^{\ at},\ a > 0$	
	$te^{\ at},\ a > 0$	
	$te^{-at},\ a > 0$	
	$t^n e^{\ at},\ n = 1, 2, 3, ...$	
	$\frac{1}{(n-1)!} t^{n-1} e^{at}$	
	$\frac{1}{\Gamma(k)} t^{k-1} e^{at},\ k > 0$	
cosine	$\cos\!t$	
sine	$\sin\!t$	
	$e^{\ at} \cos\!t\ e^{\ at} \sin\!t\ t\cos\!t\ t\sin\!t$	
hyperbolic cosine	$\cosh\!t$	$\frac{s}{s^2-\omega^2} s^2 - \omega^2 \neq 0$
hyperbolic sine	$\sinh\!t$	$\frac{\omega}{s^2-\omega^2} s^2 - \omega^2 \neq 0$

rem lim $f(t)$ = lim $sF(s)$ connecting the $t!0$ $s!1$ Laplace and inverse Laplace transforms. Again, using the Laplace transform formula (5.9), the Laplace-transformed expressions for selected differential, integral, and convolution functions are as given in Table 5.4.

Table 5.4: Differential, integral, and convolution functions.

1st		$sF(s)$ $f(0)$
2nd		$s^2F(s)$ $sf(0)$ $f^0(0)$
3th		$s^3F(s)$ $s^2f(0)$ $sf^0(0)$ $f^{00}(0)$
	$\geq$	$F(s) = L(f(t))$, $s > 0$
– derivative		–
– derivative		– –
– derivative		– – –
		$s^nF(s) - s^{n\,1}f(0) -$
nth derivative	$f^{(n)}(t)$	$s^{n2}f^0(0)$ :::::::: $f(n1(0)$
multiplication	$tf(t)$	$F^0(s); F(s) = L(f(t))$
multiplication	$tf(t)$	$F^0(s); F(s) = L(f(t))$
multiplication	$t^2f(t)$	$F''(s) = \frac{d^2F(s)}{ds^2}$
multiplication	$t^nf(t)$	(1) $^nF^{(n)}(s)$, $F^{(n)}(s) = \frac{d^nF}{ds^n}$
multiplication	$te^{at}; a > 0$	$-\frac{d}{ds}\left(\frac{1}{s-a}\right) = -\frac{1}{(s-a)^2}$
multiplication	$te^{at}; a > 0$	$\frac{d}{ds}\left(\frac{1}{s+a}\right) = -\frac{1}{(s+a)^2}$
multiplication	$te^{at}; a > 0$	–
multiplication	$t\cos! t$	$\frac{d}{ds}\left(\frac{s^2 \; \omega^2}{(s^2+\omega^2)^2}\right)$
multiplication	$t\sin! t$	$\frac{d}{ds}\left(\frac{2s\omega}{(s^2+\omega^2)^2}\right)$
integration	$Rf(t)dt$	$\frac{F(s)}{s} - \frac{f(0)}{s}$
convolution	$(f*g)(t) = \int_0^t f(\tau)g(t-d)$ $= \int_0^t f(t-\tau)g(\tau)d\tau$	$L(f\,g) = L(f)L(g)$

Time Functions Frequency Typical Functions $f = f(t)$, $t \geq 0$ Functions
$f'(t)$
$f''(t)$
$f'''(t)$

For the armature current $i_a(t)$ of the servomotor, the voltage drop v_{Ra} across the armature resistance R_a is described by $vR_a(t) = R_a \frac{dq(t)}{dt}$ and its Laplace transform is L $(v_{Ra}(t)) = R_a(sQ(s) \, q(t_0)$, where $q(t_0)$ is the initial electric charge at time $t = t_0$. The Laplace transform of the voltage drop at the inductance $vL_a(t) = L_a \frac{d^2q(t)}{dt^2}$ is $L(v_{La}(t)) = L_a$ $(s^2Q(s) \, sq(t_0) \, q\_(t_0))$, where $q\_(t_0)$ is the initial current in the electric circuit of the servomotor. The inverse Laplace transform of the circuit current $I_a(s) = sQ(s) \, q(t_0)$ of the servomotor is $\mathcal{L}^{-1}(I_a(s)) = \frac{dq(t)}{dt}$. The expressions in Tables 5.2–5.4 of Laplace transforms are helpful in solving ODEs involving constant coefficients, first-order, second-order, or higher order derivative functions.

5.5 Results and discussion

5.5.1 Introduction

This section is an accumulation of the data collected and the results of the calculations and tests performed throughout the project. We start with the data collected from measuring the components of the Stirling engine kit and the resulting SolidWorks drawings. From there we provide the results of the tests performed relating to the different temperatures and environments the Stirling engine was tested in. Next, we include the data collected from testing the revolutions per minute of the flywheel. We then review the results of the tests performed on the e-waste materials.

5.5.2 Guidelines for selecting and using e-waste materials

The guidelines for selecting and replacing the components of the Stirling engine with e-waste materials are drawn from our engineering design iterations, testing, and analysis. Admittedly, these guidelines will reduce the manual effort and cost burden of the unstructured ways of managing e-waste in Ghana. The co-design activities and interviews with the students and professors at ACUC in Ghana further suggested that the proposed guidelines will bring about a structural shift for an increased e-waste economy. The voltage generated by the Stirling engine with an e-waste servomotor in the assembly was between 4.8 and 5 V. Our laboratory tests and analysis quantified this required voltage for charging cell phone in Ghana's remote places. The e-waste site in Agbogbloshie, Ghana, is consistent with what is seen in many other e-waste sites in sub-Saharan Africa. There are several factors that account for the large variations of e-waste in Ghana and other African countries. From political incentives to business opportunities for more productive uses of e-waste are some of the factors driving the quantities, weight, and variety of e-waste in communities across in Ghana. Interviews with our partners in Ghana – combined with research and analysis – indicated the variety of e-waste, especially computers, laptops, televisions, mobile phones, printers, and many other bulky electronics available in the country. The proposed guidelines, as described in the MQP report, will provide a background of greater awareness about more productive uses of e-waste, toxic substances that harm health and the environment, and safety practices for e-waste workers. It is conceivable to use the guidelines to create e-waste trade codes, strengthen e-waste regulations, and practical methods for disposing end-of-life e-waste products.

5.5.3 Initial component measurements and drawings

The team began collecting data on the original Stirling engine to better understand the components involved in the engine. This includes data collection for each individual part in the engine and is presented as a bill of materials and the dimensions of the parts, which can be found in Appendices B and C, respectively. The engine was disassembled, so the mass of each individual part could be measured. From there, the mass of each subassembly and the full Stirling engine was found. The team performed a series of tests on the original Stirling engine to assess the functionality and efficiency of the engine. The purchased Stirling engine was advertised with the ability to reach an output voltage range of 5–9 V. When the heat was first applied to the hot cylinder of the engine, the highest output voltage reached was approximately 2.6 V. The team then adjusted the length of the piston arm on the cold cylinder of the engine. The length was changed by adjusting where the pin of the flywheel was attached to the piston arm as shown in Figure 5.17.

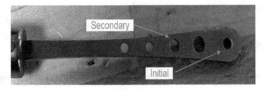

Figure 5.17: Piston arm with indicated length adjustment.

The initial and secondary holes for the adjustments are depicted in this figure. We started with adjusting it by one hole difference, which was approximately 0.18 in, leading to an increase of around 1 V. The engine was running at about 3.6 V after this change. Unfortunately, the next hole on the piston arm was not cut large enough for the pin, so the team instead adjusted the length of the arm on the hot-cylinder piston. The adjustment to the piston in the connector decreased the volume in the hot cylinder, which increased the voltage. This increase was due to the rise in pressure in the hot cylinder that resulted in a greater difference between the hot and cold cylinder pressures. This adjustment also decreased the fluid in the hot cylinder that needed to be heated up, which allowed it to heat faster, therefore, increasing the output speed of the flywheels, in turn boosting the output voltage.

The new engine that was purchased produced 6 V of output. When the engine reached this voltage, the team attached a cell phone to see the results. The phone recognized the voltage and indicated that it began to charge. Shortly after, the resistance caused by the phone made the voltage output drop to 3.8 V. The phone then stopped charging. To keep the cell phone charging, the Stirling engine must maintain a 5 V reading or higher when the phone is connected. Due to the drop in voltage from the resistance, we were unable to keep the cell phone charging consistently with the open flame as a heat source. From a suggestion made by our advisor, we performed a few tests on the engine with a more direct heat source: a torch. This was to test if the engine could produce a higher voltage with a consistent heat source. The torch was lit and held about 8–9 in away from the hot cylinder, and with a gentle push of the flywheel, the engine started running. It reached its maximum voltage within less than a minute, faster than all other trials, and had the highest reading out of any previous runs of 7.4 V. We connected a cell phone and it began charging. Even with the resistance the phone continued its charge.

5.5.4 Temperature and environment testing

To test the Stirling engine's ability to charge a cell phone or a battery to charge a cell phone, the team conducted a series of tests and calculations. The team identified that the surrounding temperature of the Stirling engine could have significant effects on the voltage output and the engine's efficiency.

The team tested the following factors at different temperatures:

Maximum voltage

Time for the engine to start after igniting the flame

Time to reach maximum voltage

Time to stop after extinguishing the flame

Figures 5.17–5.19, and 5.21 are graphs of the results for these measurements from the original engine. Additional information can be found in Appendix C. The graph in Figure 5.18 shows that the points are scattered and mostly concentrated on a region for the temperature > 70 °F.

We have a similar situation in the subsequent graphs in Figures 5.19 and 5.20.

These tests indicate that the room temperature the engine runs in can affect the efficiency of the engine. When the temperature was run in very cool temperatures of 49.5 °F, the time to start the engine was significantly higher than when the temperature was 70 °F and higher. We believe this is because the hot cylinder is cooler at the start, causing it to take longer to reach a temperature hot enough to create the pressure difference that allows for the engine to run. There does not appear to be any significant change in efficiency of the engine when the temperature only differs by a

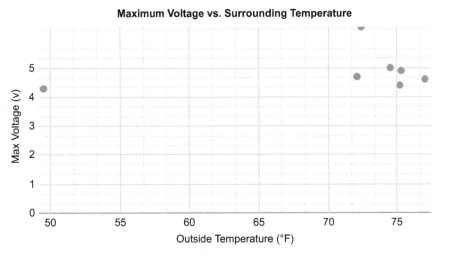

Figure 5.18: Graph of maximum voltage versus surrounding temperature.

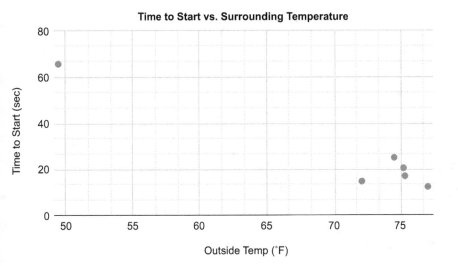

Figure 5.19: Graph of the time to start the engine versus differing surrounding temperature.

few degrees. Based on observation, after the engine runs for multiple iterations of the tests, the efficiency decreases. The team theorized that this decrease in efficiency correlates to the decrease in temperature difference between the two cylinders. The flame's heat that is applied to the hot cylinder is also being applied to the cold cylinder due to their proximity. The team started to investigate the application of heat sinks on the engine as pictured in Figure 5.22.

We used the Stirling engine in Figure 5.22 to conduct more tests to reach a proper conclusion, but based on preliminary tests, the heat sinks created a barrier between the

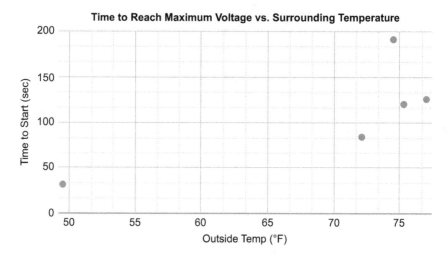

Figure 5.20: Graph of time to reach the maximum voltage versus surrounding temperature.

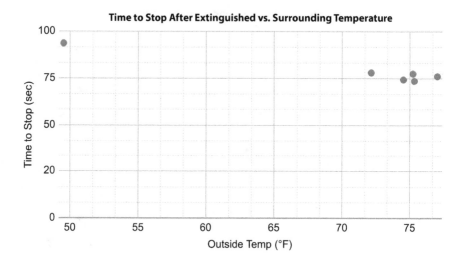

Figure 5.21: Graph of time to stop after extinguished versus differing surrounding temperature.

flame and the cold cylinder and increased the voltage output by a few decimals of voltages. Additional data from the heat sink test can be found in Appendix C.

5.5.5 Flywheel speed and waste test results

The flywheel speed was found to have inconsistent results. As shown in the graph in Figure 5.23, the results of the speed readings were drastically different for the same

Figure 5.22: Stirling engine with recycled heat sinks applied between hot and cold.

voltage outputs. This is inconsistent with the calculations of revolutions per minute correlating to voltage output, for they should have similar readings. The team believes this is due to an error with the tachometer we used that had imprecise and inconsistent readings. The result also could have been affected by the reflective material of the flywheel. The tachometer runs by shining a laser off of a reflective piece of tape; however, since the entire Stirling engine is reflective, the device was picking up inconsistent readings. After applying blackout tape to the flywheel, the readings were still inaccurate. The graph in Figure 5.23 shows the range of results. Additional data for the graph can be found in Appendix C.

With regard to the selection and replacement of the Stirling components with e-waste, the team tested the functionality of the e-waste material to identify its ability to replace parts of the engine. These examinations included functionality tests as well as efficiency tests to ensure the capabilities of the e-waste parts. The parts that we tested were motors, heat sinks, and elastic bands. The team found motors from a used CD player and a toy car. The test conducted on the motor was purely for functionality. The team replaced the motor on the original Stirling engine with the e-waste motors to assess their performance.

Each motor was individually replaced and tested to determine if they could produce an output voltage. Figure 5.24 displays the results. Additional information for the graph can be found in Appendix C.

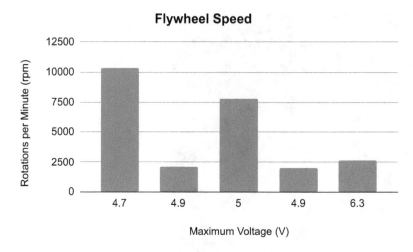

Figure 5.23: Graph of the rotations per minute versus maximum voltage.

The output voltage produced by the recycled motors reached a maximum of about half the output voltage required to charge a cell phone. Due to a malfunction during the testing of the toy car engine the team observed, further study and calculations were needed. While testing the toy car engine, it took significantly longer for the engine to reach the necessary starting temperature, and when the engine did start, it had a much lower voltage output than any of the previously tested motors. The team assumed that the low voltage was because it was an old, used engine. This problem was solved during one test when the belt attaching the cold cylinder flywheel and the

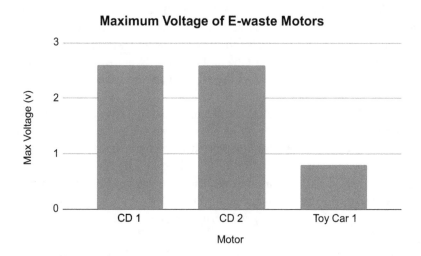

Figure 5.24: The maximum voltage of different e-waste motors.

motor flywheel became misaligned and slipped off. When this occurred, the engine significantly increased in speed. This was due to the toy car motor having a much higher resistance than that of any other motor the team had previously tested.

5.5.6 Servomotor and Stirling engine performance analysis

We want to find values for the circuit elements R_a, C_a, L_a of the servomotor for which the required voltage to charge a cell phone is produced. In addition to locating specific values for the parameters R_a, C_a, and L_a, the time and speed required to produce a voltage near or equal to 5 V are also of interest for replacing the Stirling engine components with e-waste materials. The time and speed for producing the current and voltage for the proposed charging system are determined by supplying a constant heat source to the Stirling engine. Taking term-by-term Laplace transform of the servomotor equation (5.8) and simplifying the algebra led to the electric charge

$$Q(s) = \frac{(s + 2\xi\omega_n)q_0 + \upsilon_0}{s^2 + 2\xi\omega_n s + \omega_n^2} + \frac{\omega_n R(s)}{s^2 + 2\xi\omega_n s + \omega_n^2}, \quad R(s) = \frac{1}{\kappa_C}\mathcal{L}(f(t)), \tag{5.11}$$

which is now a function of the Laplace transform variable s and $L(f(t)) = L(v_a(t)\ v_e(\_)$ $v_c(t_0))$. The quantities q_0 and v_0 are the initial values in the servomotor. In this servomotor equation, $\mathcal{L}(q(t)) = Q(s)$ and $\mathcal{L}(r(t)) = \frac{1}{\kappa_C}\mathcal{L}(f(t)) = R(s)$. On the left-hand side of the electric charge function $Q(s)$, the rational expression containing the initial conditions q_0 and v_0 is referred to as the complementary function or transient function. For given initial values for q_0 and v_0, the growth and decay rates of the armature current $i_a(t)$ of the servomotor can be determined as the damping ratio $= \frac{R_a}{2\omega_n L_a}$ and undamped natural frequency $!^n = \sqrt{\frac{\kappa_C}{L_a}}$ are varied. The growth and decay rates are separated from each other when damping coefficient $c = R_a$ and critical damping $c_{\mathrm{crit}} = 2pL_a$ $c = 2L_a!_n$ are identical. The rational expression containing the applied voltage $R(s)$ in equation (5.11) is called the particular function or the steady-state function. The inverse Laplace transform of both the transient and steady-state functions of $Q(s)$ gives the electric charge $q(t)$, the armature current $i_a(t)$, and corresponding voltage drops $v_{La}(t)$ for the inductance L_a, $v_{Ra}(t)$ for the resistance R_a, and $v_{Ca}(t)$ for the capacitance C_a.

5.5.7 Characteristic equation and poles of the servomotor

Applying the Laplace transform method to the servomotor equation (5.8) provides frequency function $Q(s)$ for the electric charge as given in equation (5.11). This frequency function of $Q(s)$ is a function of the Laplace transform variable s and it is composed of the circuit elements R_a, C_a, L_a. The transient and steady-state frequency functions for the electric charge $Q(s)$ are infinity for values of s satisfying the characteristic equation

$$D(s) = s^2 + 2\xi\omega_n s + \omega_n^2 = 0, \tag{5.12}$$

where $= \frac{R_a}{2\omega_n L_a}$, $\omega_n = \sqrt{\frac{\kappa_C}{L_a}}$, and $\kappa_C = \frac{1}{C_a}$, $C_a{\neq}0$. Roots and poles are typical names for the values of s satisfying the characteristic equation $D(s) = 0$. Using the well-known quadratic formula for solving the standard quadratic equation $ax^2 + bx + c = 0$, we obtain the poles

$$s_1 = !_n + !_n q^2 1, \quad s_2 = !_n \ !_n q^2 \ 1 \tag{5.13}$$

of the servomotor for the discriminant $=^2 1 > 0$. The poles s_1 and s_2 are located on the real line in the left-hand side of the complex plane. In the complex plane, the vertical axis is designated as the imaginary axis and the horizontal axis as the real axis. The complex index $j = p\ 1$ is marked on the vertical axis of the complex plane. Denoting the discriminant by $=^2 1$, maintaining the undamped frequency $!^n = \sqrt{\frac{\kappa_C}{L_a}}$ as a constant and varying the damping ratio $= \frac{R_a}{2\omega_n L_a}$, the poles of the characteristic equation $D(s) = 0$ will change. In essence, as the discriminant $=^2 1$ changes, the poles of the characteristic equations $D(s) = 0$ can be real and distinct poles, real and repeated poles, or occur in complex conjugate poles of the form $s_{1,2} = j!$ where the real part $\text{Re}(s_{1,2}) =$ and imaginary part $\text{Im}(s_{1,2}) = !$.

5.5.8 Servomotor and Stirling engine performance

The complex conjugate poles $s_{1,2} = j!$ with $= !_n$ and $! = !_n p1^2$ when discriminant $=^2 1 < 0$. For any value of $= \frac{R_a}{2\omega_n L_a}$ in the open interval (0,1) of the real line $R = (1,1)$, the performance of the servomotor is asymptotically stable by Lyapunov sense, and the Stirling engine generates more armature current than the case when $R_a > 2!_n L_a$. Concrete values for the armature resistance R_a, armature inductance L_a, and armature capacitance C_a of the servomotor of the Stirling engine can be selected in such a way that the damping ratio satisfies the inequality $0 < < 1$ and for a fixed undamped natural frequency $!^n = \sqrt{\frac{\kappa_C}{L_a}}$. As the time t approaches to infinity, the armature current $i_a(t)$ decays exponentially. The torque driving the moment of inertia of the loads on the servomotor output shaft will also decay exponentially as $t\ !1$.

When the discriminant $=^2 1 = 0$, the damping ratio $= \frac{R_a}{2\omega_n L_a} = 1$. This is referred to as the critical case that distinguishes regions of slow and fast performance of the servomotor and the Stirling engine. The poles s_1 and s_2 of the characteristic equation $D(s) = 0$ are real and equal. Incidentally, the values of the armature resistance R_a and inductance L_a of the servomotor are also equivalent. When the armature resistance

R_a of the servomotor is zero, the regions of stable and unstable performance of the servomotor of the Stirling engine are established. In the unstable region, the armature current $i_a(t)$ and torque $\tau_m = k_t i_a(t)$ of the servomotor will grow exponentially as $t \to 1$, while in the stable region, these physical quantities will decay exponentially as $t \to 1$. In the unstable region, the charging of cell phones with the current generated by the Stirling engine will experience interruptions. There will be also persistent breakdowns in the flow of the armature current $i_a(t)$ into the cell phone chargers. The problem causing the instability in the current flow into the cell phone chargers should be fixed first and then resume the charging of the cell phones using the Stirling engine. For the discriminant $= \xi^2 1 > 0$, the damping ratio $= \frac{R_a}{2\omega_n L_a} > 1$ and the poles of the characteristic equation $D(s) = 0$ are $s_1 = \xi_n + \xi_n p^2 1$ and $s_2 = \xi_n \xi_n p^2 1$. This is the case when concrete values for the armature inductance L_a must satisfy the inequality $0 < L_a < \frac{R_a}{2\omega_n}$ and with $\xi_m = \sqrt{\frac{K_C}{L_a}} \neq 0$. Although both the armature current $i_a(t)$ and torque $\tau_m = k_t i_a(t)$ of the servomotor will decay exponentially as the time $t \to 1$, the performance of the Stirling engine is slower than the case when the damping ratio satisfies the inequality $0 < \frac{R_a}{2\omega_n L_a} < 1$. Setting the voltage difference $v_a(t) \; v_e() v_{ca}(t_0) = 0$ will yield $R(s) = 0$ in equation (5.11) and the expression for transient electric charge

$$
\begin{aligned}
Q_{trs}(s) &= \frac{(s + \xi\omega_n)q_0 + v_0 + \xi\omega_n q_0}{s^2 + 2\xi\omega_n s + \omega_n^2} \\[2mm]
&= \frac{(s + \xi\omega_n)q_0}{s^2 + 2\xi\omega_n s + \omega_n^2} + \frac{v_0 + \xi\omega_n q_0}{s^2 + 2\xi\omega_n s + \omega_n^2} \\[2mm]
&= \frac{(s + \xi\omega_n)q_0}{(s + \xi\omega_n)^2 + \omega_n^2(1 - \xi^2)} + \frac{v_0 + \xi\omega_n q_0}{(s + \xi\omega_n)^2 + \omega_n^2(1 - \xi^2)}
\end{aligned}
\tag{5.14}
$$

for $0 < \frac{R_a}{2\omega_n L_a} < 1$. This transient equation (5.14) for the electric charge is obtained after performing several algebraic manipulations to find a match in the Laplace transform Tables 5.2–5.4. The expression $\omega_d = \xi_n p1^2$ is referred to as the damped natural frequency and nondimensional frequency ratio for the case $0 < \frac{R_a}{2\omega_n L_a} < 1$ is $\frac{\omega_d}{\omega_n} = \sqrt{1 - \xi^2}$, where $\xi = \frac{R_a}{2\omega_n L_a}$ and $\xi_m = \sqrt{\frac{K_C}{L_a}}$. Taking the inverse Laplace transform of the transient electric charge $Q_{trs}(s)$ in equation (5.14) and simplifying lead to the electric charge as a function of time

$$q_{\text{trs}}(t) = ae^{-\xi\omega nt}\left(\frac{C_1}{\sqrt{C_1^2 + C_2^2}}\cos\omega_d t + \frac{C_2}{\sqrt{C_1^2 + C_2^2}}\sin\omega_d t\right), \quad \omega_d = \omega_n\sqrt{1-\xi^2}$$

$$= ae^{-\xi\omega nt}(\cos\omega_d t\cos\varphi + \sin\omega_d t\sin\varphi), \quad C_1 = q_0, \quad C_2 = \frac{\upsilon_0 + \xi\omega_n q_0}{\omega_d} \tag{5.15}$$

$$= ae^{-\xi\omega nt}\cos(\omega_d t - \varphi), \quad a = \sqrt{C_1^2 + C_2^2}, \quad \varphi = \tan^{-1}\left(\frac{C_2}{C_1}\right)$$

where the constants C_1 and C_2 are determined using the initial conditions $q(t_0) = q_0$ and $q\_(t_0) = v_0$. The speed at which the Stirling engine produces the transient current and servomotor torque is determined by the reciprocal of the product of the damping ratio and undamped natural frequency $!_n$. Differentiating the expression of the transient electric charge $q_{\text{trs}}(t)$ with respect to time gives the transient armature current produced by the Stirling engine. Using the current–torque relationship $m = k_t i_a(t)$, an explicit expression for the transient torque is also obtained. Selecting and replacing e-waste materials for the Stirling engine component can still produce the voltage required for charging a cell phone if the values for the resistance R_a, inductance L_a, and capacitance C are consistent with values in the interval $0 < \frac{R_a}{2\omega_n L_a} < 1$ and undamped natural frequency $!^n = \sqrt{\frac{\kappa c}{L_a}}$ with $c = 1 = C_a$.

To provide additional guidelines for selecting unused and e-waste component materials to design and build charging systems for Ghana's remote places, we let a represent any positive constant voltage produced by the Stirling engine. The Laplace transform of the constant voltage $r(t) = a$ is $R(s) = \mathcal{L}(r(t) = \frac{a}{s}$, where the Laplace variable $s = 06$. Substituting this expression $R(s) = a = s$ into the function of the electric charge $Q(s)$ in equation (5.11), we obtain the steady-state expression of the electric charge

$$Q_{\text{ss}}(s) = a\frac{\omega_n^2}{s(s^2 + 2\xi\omega_n s + \omega_n^2)} = \frac{ak_1}{s} + \frac{ak_2 s + ak_3}{s^2 + 2\xi\omega_n s + \omega_n^2}$$

$$= a\left(\frac{1}{s} - \frac{s + 2\xi\omega_n}{s^2 + 2\xi\omega_n s + \omega_n^2}\right), \quad k_1 = 1, \quad k_2 = -1, \quad k_3 = -2\xi\omega_n \tag{5.16}$$

$$= a\left(\frac{1}{s} - \frac{s + 2\xi\omega_n}{(s + \xi\omega_n)^2 + \omega_n^2(1-\xi^2)}\right)$$

for any positive constant voltage a produced by the Stirling engine and the initial conditions $q_0 = v_0 = 0$. This steady-state expression $Q_{\text{ss}}(s)$ of the electric charge is due to the constant voltage a produced by the Stirling engine. In equation (5.16), the constants k_1 and k_2 are the partial fraction coefficients. The damping ratio lies in the interval $0 < \frac{R_a}{2\omega_n L_a} < 1$ and undamped natural frequency $!^n = \sqrt{\frac{\kappa_C}{L_a}}$ with $c = 1 = C_a$. Taking the inverse Laplace transform of equation (5.16) and simplifying the algebra, we obtain the electric charge

$$
q_{ss}(t) = a\mathcal{L}^{-1}\left(\frac{1}{s} - \frac{s + \xi\omega_n}{(s + \xi\omega_n)^2 + \left(\omega_n\sqrt{1-\xi^2}\right)^2} - \frac{\xi}{\sqrt{1-\xi^2}}\left(\frac{\omega_n\sqrt{1-\xi^2}}{(s + \xi\omega_n)^2 + \left(\omega_n\sqrt{1-\xi^2}\right)^2} \right) \right)
$$

$$
= a - ae^{-\xi\omega_n t}\left(\cos\omega_d t + \frac{\xi}{\sqrt{1-\xi^2}}\sin\omega_d t \right)
$$

$$
= a - \frac{ae^{-\xi\omega_n t}}{\sqrt{1-\xi^2}}\left(\sqrt{1-\xi^2}\cos\omega_d t + \xi\sin\omega_d t \right),\ \omega_d = \omega_n\sqrt{1-\xi^2}
$$

$$
(5.17)
$$

as a function of time. Differentiating $q_{ss}(t)$ with respect to time gives the steady-state armature current $i_a(t)$ and substituting the result from the differentiation into the current torque equation $m = k_t i_a(t)$ provides the steady-state expressions for the current and torque, respectively. The heat source, Stirling engine cycles, and power transmission from one component to another in the Stirling engine should be done in such a manner that selected values for the resistance R_a, capacitance C_a, and inductance L_a equations must satisfy the expressions in equations (5.12), (5.15), and (5.17) over the interval $0 < \frac{R_a}{2\omega_n L_a} < 1$. These three equations guide the selection of component materials for the Stirling, servomotor, power-transmitting mechanism, and the cell phone charging system. Experimentally, the team could show that the voltage produced by the Stirling engine for charging cell phones in remote places in Ghana was between 4.8 and 5 V. In the interval (i) $0 < \frac{R_a}{2\omega_n L_a} < 1$, the Stirling engine would generate the voltage for charging cell phones faster than operating the Stirling engine over the intervals (ii) $\frac{R_a}{2\omega_n L_a} = 1$ and (iii) $\frac{R_a}{2\omega_n L_a} > 1$. Both cases tend to be susceptible to greater heat loss than the case for $0 < \frac{R_a}{2\omega_n L_a} < 1$. Our testing and experimental observations further showed that the insulation of the heat source can prevent the escape of a large amount of heat energy to the surroundings of the Stirling engine. The preferred operating conditions for the Stirling engine to generate sufficient voltage for charging cell phones in Ghana's public places are when $0 < \frac{R_a}{2\omega_n L_a} < 1$ and $R_a = 0$ with the undamped natural frequency being! $!^n = \sqrt{\frac{\kappa_C}{L_a}}, \kappa_C = 1/C_a$.

5.5.9 Stirling engine and cell phone charging

The most common way to charge mobile cell phones is plugging the USB cable of the cell phone into a power supply outlet. Incompatible cell phone chargers or fluctuating current flow through the USB cable could damage the cell phone. Host computer, laptop, and other modern electronic devices are alternative ways of charging cell phones through the USB cable. Wireless power transfer, inductive charging coils, and electric energy sources enable cell phones to continuously charge. Worldwide, there are growing expectations to have access to charging systems for cell phones. At conferences, trade shows, and mass-gathering public places, people want to obtain access to cell phone charging systems. The Stirling engine, servomotor, and the voltage—current battery constitute our proposed charging system for the people of Ghana. The flywheel, piston–crankshaft, pulleys, belts, bearings, servomotor, and other power-transmitting mechanisms in the Stirling engine assembly convert the thermomechanical energy into current, voltage, and torque. Our test and calculation results showed that the Stirling engine can produce 5 V. As the intensity of the applied heat source to the Stirling engine changes, we found the generated voltage to vary between 4.5 and 4.8 V. The adjustments of the pins on the piston–crankshaft, friction between the piston and cylinder, and other tolerances supporting the performance of the Stirling engine improved the resulting output voltage.

The performance of the proposed charging system for cell phones is analyzed theoretically by the exponential growth and decay rates of the armature current $i_a(t)$, voltage drops $v_{La}(t)$ across the inductance L_a, $v_{Ra}(t)$ across the resistance R_a, and $v_{Ca}(t)$ across the capacitance C_a, and the servomotor torque $m = k_t i_a(t)$. For the Stirling engine operating conditions satisfying the inequalities (i) $0 < \frac{R_a}{2\omega_n L_a} < 1$, (ii) $\frac{R_a}{2\omega_n L_a} = 1$, and (iii) $\frac{R_a}{2\omega_n L_a} > 1$ with fixed undamped natural frequency $!^n = \sqrt{\frac{\kappa_C}{L_a}}, L_a \neq 0$, the current $i_a(t)$, voltage drops v_{La} (t), $v_{Ra}(t)$, $v_{Ca}(t)$, and torque $m = k_t i_a(t)$ decay exponentially as the time t !1. These three inequalities guide the selection of concrete values for the resistance R_a, inductance L_a, capacitance C_a, and the amount of the applied heat source on the Stirling engine. The operation of the Stirling engine for values of the servomotor elements in the inequality $0 < \frac{R_a}{2\omega_n L_a} < 1$ will produce sufficient voltage for the charging system. In this case, the performance of both the Stirling engine and servomotor is better than the cases when $\frac{R_a}{2\omega_n L_a} = 1$ and $\frac{R_a}{2\omega_n L_a} > 1$. The Stirling engine produced a voltage for our charging system consistent with the same amount of voltage produced by smart devices, host computer, laptop, and power supply outlet.

5.6 Conclusion and future recommendations

Our goal of the project was to design a Stirling engine out of e-waste materials, but due to lack of time, resources, and working on the project remotely we were unable to reach this goal. The official length of time we intended to work on the project was 7 weeks, but we quickly realized that this was not enough. Halfway through the timeframe, we had to adjust our approach and modify the objectives – still hoping to achieve our goal. From week 1, we could not begin our first or second objective until we received the Stirling engine in the mail, which took more than a week. A large factor that also caused challenges was not being present in Ghana. The 5-h time difference and lacking direct access to Agbogbloshie made it difficult to gather all the information we wanted. Furthermore, when we met with the ACUC team, it was productive and informative but coordinating schedules and finding meeting times proved to be difficult. If we were in person with the ACUC students and e-waste site workers, our team believes that the project would have been more successful. Although our project changed throughout the process, we kept many aspects in mind that we will provide as recommendations for future work. These suggestions are further discussed in Section 5.2.

The following section discusses the collaborative experience with the Ghanaian community and future proposals to improve the project. Throughout this undertaking, a prominent facet was our collaboration with the community in Ghana.

Co-design holds an emphasis on the relationships among producer, designer, and consumer, the responsibilities that come with each, how they overlap, and how the positions work together. Co-design is important to our project as we strive to ensure the sustainability and longevity of the work we accomplish. By using an approach that embodies the ideals of co-design, all the knowledge that our MQP team learns over the term will be simultaneously shared with our partners in Ghana and vice versa. This means we are designing collectively and after our MQP has ended, the team in Ghana can continue building the project. Implementing co-design in the project allowed the ACUC students to continue working on substituting components and redesigning the Stirling engine out of e-waste parts.

For the future of this project or projects like it, the team has suggestions based on the work and observations made throughout the process. One aspect of our testing that became an obstacle was how to run the engine without using a direct flame. The engine runs with a flame under the hot cylinder. Heat can be dangerous in any mechanism, and too much of it could lead to injuries, explosions, or burns if something goes wrong. The team had to ensure our testing was safe and that we knew how to mitigate an emergency. When testing the engine, we wore safety goggles, maintained an appropriate distance from the flame, and always had tongs and water to safely extinguish the fire if anything went wrong. We also made sure to conduct the testing in a laboratory with the necessary fire safety equipment. The safety precautions we had to take are not sustainable for a household use product. An improvement that may be applied to the engine is an alternative heat source that does not include an open flame. This will improve the safety of the engine as well as the possibility of increasing the efficiency of the engine depending on the source. To improve the design and create a sustainable Stirling engine design, one option is to use solar power instead of fire. Also, the use of renewable energy sources, such as solar power, would minimize the pollutants compared to that of burning isopropyl alcohol.

One project goal was to replace the original components on the engine with recycled materials. Reusing and recycling materials is good for the environment as it can create new products without producing additional waste. Without this type of recycling, waste at e-waste sites, like Agbogbloshie, would never get reused and this could lead to an accumulation of waste. This buildup would likely take up more space than there is available, and it could cause harm to the environment and to the health of those near it. To continue the design process, more components can be substituted with e-waste. The project focused on structural component replacement, but with more time, machining can be implemented on recycled parts to reach the high tolerances we need on the mechanical parts of the engine. The goal here is to build an engine from scratch using only e-waste materials. This engine could be replicated by others and then sold for profit. Charging a phone is valuable to the Ghanaian community for the reasons we previously discussed. This engine could also be used to charge a battery pack that would have more versatile capabilities; it could charge other electronics such as computers or headphones. For a future design, the engine can be devised to produce a voltage high enough to sustain a charge into a battery pack.

References

[1]   Ritchie, H. & Roser, M. (2020, November 28). Access to Energy. Our World in Data. Retrieved January 27, 2022, from https://ourworldindata.org/energy-access

[2]   Owusu-Adjapong, E. (2018, April 9). Dumsor: Energy Crisis in Ghana. Stanford University. https://energyrights.info/sites/default/les/artifacts/media/pdf/dumsor_energy_crisis_in_ghana.pdf

[3]   Kumi, E. N. (2017) The Electricity Situation in Ghana: Challenges and Opportunities. Washington, DC, Center for Global Development, 2017, from https://www.cgdev.org/sites/default/les/electricity-situation-ghana-challenges-and-opportunities.pdf.

[4]   Valickova, P. & Elms, N. (2021). The Costs of Providing Access to Electricity in Selected Countries in Sub-Saharan Africa and Policy Implications. *Energy Policy, 148*(Part A). https://doi.org/10.1016/j.enpol.2020.111935.

[5]   Panos, E., Densing, M., & Volkart, K. (2016). Access to Electricity in the World Energy Council s Global Energy Scenarios: An Outlook for Developing Regions until 2030. *Energy Strategy Reviews, 9*, 28–49. https://doi.org/10.1016/j.esr.2015.11.003. Retrieved from https://www.sciencedirect.com/science/article/pii/S2211467X15000450

[6]   Eshun, M. E. & Amoako-tuffour, J. (2016). A Review of the Trends in Ghana s Power Sector. *Energy, Sustainability and Society, 6*(1). https://doi.org/10.1186/s13705016-0075-y.

[7]   Nduhuura, P., Garschagen, M., & Zerga, A. (2021, June 20). Impacts of Electricity Outages in Urban Households in Developing Countries: A Case of Accra, Ghana. MDPI. Retrieved January 20, 2022, from https://www.mdpi.com/1996-1073/14/12/3676/htm

[8]   Arku, G., Luginaah, I., & Mkandawire, P. (2012). You Either Pay More Advance Rent or You Move Out: Landlords/Ladies and Tenants Dilemmas in the Low-income Housing Market in Accra. *Ghana Urban Studies, 49*(14), 3177–3193. https://doi.org/10.1177/0042098012437748.

[9]   Enerdata (2020) Ghana Energy Information. Enerdata, from https://www.enerdata.net/estore/en
      ergy-market/ghana/#:~:text=Total%20Energy%20Consumption,consumption%20was%20485%
      20kWh%2Fcap.
[10] Quartey, J. D. & Ametorwotia, W. D. (2017). Assessing the total economic value of electricity in
      Ghana: A step toward energizing economic growth. Retrieved from https://www.theigc.org/wp-
      content/uploads/2018/06/Quartey-Ametorwotia-2017Final-report.pdf.
[11] Spitzbart, M. (2022, February 10). Environmentally sound disposal and recycling of e-waste in
      Ghana. Giz. Retrieved February 11, 2022, from https://www.giz.de/en/worldwide/63039.html#:~:
      text=In%20Ghana%2C%2095%20per%20centrecycling%20are%20organised%20largely%20infor
      mally.&text=The%20most%20important%20location%20is,commonly%20known%20as%
      20Agbogbloshie
[12] Lepawsky, J. & McNabb, C. (2010). Mapping International OWS of Electronic Waste. *The Canadian
      Geographer / Le Gøographe Canadien, 54*, 177–195. https://doi.org/10.1111/j.15410064.2009.00279.x.
[13] Daum, K., Stoler, J., & Grant, R. (2017). Toward a more Sustainable Trajectory for E-waste Policy: A
      Review of a Decade of E-waste Research in Accra. *Ghana International Journal of Environmental
      Research and Public Health, 14*(2), 135. https://doi.org/10.3390/ijerph14020135.
[14] Kuma, P. (2011). Old Fadama – A Community Under Threat. [Web post]. Retrieved from
      https://philipkumah.wordpress.com/page/3/
[15] Cassels, S., Jenness, S., Biney, A., Ampofo, W., & Dodoo, N. (2014). Migration, Sexual Networks, and
      HIV in Agbogbloshie, Ghana. *Demographic Research, 31*(28), 821–888. https://www.demographic-
      research.org/volumes/vol31/28/31-28.pdf.
[16] Larmer, B. (2018, July 5). E-waste Offers an Economic Opportunity as Well as Toxicity. The New York
      Times Magazine. Retrieved March 3, 2022 from https://www.nytimes.com/2018/07/05/magazine/E-
      waste-offers-an-economic-opportunity-as-well-as-toxicity.html#:~:text=In%20the%20United%
      20States%2C%20which,two%2Dthirds%20of%20heavy%20metals
[17] Grant, R. & Oteng-Ababio, M. (2013, May 16). Mapping the Invisible and Real "African" Economy:
      Urban E-waste Circuitry. *Urban Geology, 33*(1). https://doi.org/10.2747/0272-3638.33.1.1.
[18] Church, A., Greenbaum, B., & Stirling, C., (n.d.) Stirling Engine Fabrication and Design. Worcester
      Polytechnic Institute. Retrieved March 1, 2022, from https://digital.wpi.edu/show/b8515q00r
[19] Brahambhatt, R. (2021, July 28). The Everlasting, Nearly Emission Free Stirling Engine. Interesting
      Engineering. Retrieved December 14, 2021, from https://interestingengineering.com/the-everlasting
      -nearly-emission-free-stirling-engine
[20] Linda Hall Library. (2019, October 25). Scientist of the Day – Robert Stirling. Linda Hall Library.
      Retrieved March 3, 2022 from https://www.lindahall.org/robert-stirling/
[21] Walker, G. (1973). The Stirling Engine. *Scientific American, 229*(2), 80–87. http://www.jstor.org/stable/
      24923172.
[22] Denno, J. (n.d.). Design and analysis of Stirling engines – pcs.cnu.edu. Retrieved December 19, 2021,
      from https://www.pcs.cnu.edu/~dgore/Capstone/les/DennoJ.pdf
[23] Urieli, I. (2013, March 30). Stirling engine configurations – updated 3/30/2013. Retrieved
      December 20, 2021, from https://www.ohio.edu/mechanical/stirling/engines/engines.html
[24] Chen, H., Czerniak, S., De La Cruz, E., Frankian, W., Jackson, G., Shieferaw, A., & Stewart, E. (2014,
      March 28). Design of a Stirling Engine for Electricity Generation. Worcester Polytechnic Institute.
      Retrieved December 20, 2021, from https://web.wpi.edu/Pubs/Eproject/Available/E-project-032814-
      103716/
[25] inar, C., Aksoy, F., & Erol, D. (2012, June 25). The Effect of Displacer Material on the Performance of a
      Low Temperature Differential Stirling Engine. *International Journal of Energy Research, 36*(8), 911–917.
      https://doi.org/10.1002/er.1861.
[26] Nice, K. (2021, February 9). How Stirling Engines Work. HowStuffWorks. Retrieved December 19,
      2021, from https://auto.howstuffworks.com/stirling-engine.htm

[27] Lutz, A. E., Larson, R. S., & Keller, J. O. (2002). Thermodynamic Comparison of Fuel Cells to the Carnot Cycle. *International Journal of Hydrogen Energy, 27*(10), 1103–1111. https://doi.org/10.1016/s0360-3199(02)00016-2.

[28] Dimian, A. C., Bildea, C. S., & Kiss, A. A. (2014). Applied energy integration. In *Integrated Design and Simulation of Chemical Processes*, Vol. 35, 565–598. essay, Elsevier Science.

[29] Zohuri, B. (2018). Gas Power and Air Cycles. In Physics of cryogenics: An ultralow temperature phenomenon, 331–385. Elsevier.

[30] Kongtragool, B. & Wongwises, S. (2003). A Review of Solar-powered Stirling Engines and Low Temperature Differential STIRLING ENGINES. *Renewable and Sustainable Energy Reviews, 7*(2), 131–154. https://doi.org/10.1016/s1364-0321(02)00053-9.

[31] Tavakolpour, A. R., Zomorodian, A., & Akbar Golneshan, A. (2008). Simulation, Construction and Testing of a Two-Cylinder Solar Stirling Engine Powered by an At-Plate Solar Collector Without Regenerator. *Renewable Energy, 33*(1), 77–87. https://doi.org/10.1016/j.renene.2007.03.004.

[32] National Society of Professional Engineers (NSPE). (2019). NSPE Code of Ethics for Engineers. National Society of Professional Engineers. Retrieved March 3, 2022 from https://www.nspe.org/resources/ethics/code-ethics

Nada Abojaradeh
Chapter 6
When the goals change, the process must too

6.1 Introduction

The Global North has been using neoliberal development as a tool to continue the legacy of colonization. Smart development has been identified as using information technology to empower cities and communities to provide better quality of life. Smart development benefits and consequences are disproportionately distributed within these hierarchies of power [1]. WPI's Ghana Project Center has been working to address this issue by unlearning and relearning ways of thinking and is seeking to move beyond these traditional ideologies.

Therefore, a new set of goals have been redefined around the generative justice principles of prioritizing community needs and supporting the people, culture, and resources that already exist in the villages. As part of the Ghana Project Center for 2022, teams of WPI students collaborated with their local partners to push development design beyond its current framework. Projects encompass smart development, but also other technologies. These teams used non-western design ideas including co-design and generative justice to reframe their work. However, a formal step-by-step process to guide this design had not yet been defined.

The goal of this project is to map out an equitable and improved co-design process for the Ghana Project Center that puts generative justice in its forefront. The motivation for this project is to ensure that the new process does not uphold the systems the project center has been moving away from. The results of this project will provide a framework for a recommended process and work system components to support future project teams in achieving generative justice goals.

To obtain this goal, I applied a continuous improvement approach by delving deep into previous smart village design processes to understand what works and what needs to change. First, I utilized SIPOC (suppliers, inputs, process, outputs, customers) and swimlane diagrams to map out a traditional design process. Second, through co-design and generative justice intensive research and using the SEIPS (Systems Engineering Initiative for Patient Safety) methodology and depth and breadth levers, I mapped out an initial new process. Third, exploring and analyzing the work of two 2022 Ghana Project Center teams through meetings with partners and students and using tools, including PETT (People, Environments, Tools, Tasks) scans, people maps, and outcomes matrices, the new process was improved and updated until finalized.

Nada Abojaradeh, WPI student, Industrial Engineering

https://doi.org/10.1515/9783110786231-007

The project report is organized as follows: Chapter 2 provides the background and literature review needed to understand the concepts and tools utilized in this project. A summary of smart villages, the traditional design process, WPI's Ghana Project Center, generative justice, co-design, and the industrial engineering process improvement tools are described. The methodology is presented in Chapter 3, including the specific steps used to improve the traditional design process into the new process that promotes generative justice. The results of my work are presented in Chapter 4, including analysis of data from two project teams used as case studies. Chapter 4 also presents the new improved co-design process. Lastly, Chapter 5 is a summary of the work done in this project and recommendations for future work.

6.2 Background/literature review

6.2.1 The smart village

There have been many definitions for "smart" when it comes to villages. A smart village is defined as a "community empowered by digital technology and open innovation platforms to access global markets" [2]. With around 3.4 billion people living in villages, the smart village is one of the greatest opportunities to expand into the emerging market of villages around the world [3]. Global corporations and brands see these communities as an enormous, untapped source of potential for economic growth. For these companies and industry partners to expand their markets and offer the right products and services to the villagers, they need to understand what villagers want. This allows corporations to reveal what villagers are willing to pay for.

Therefore, smart village organizations can sell themselves as initiatives put in place to solve prevalent pain points by providing technologies along with innovative business models. Ideally, smart villages are intended to eradicate poverty, enhance the happiness index of rural populations, and achieve development by empowering people for economic growth through digital technologies. In summary [3], if making life better for the people living in rural villages and raising their happiness index was the only reason for smart villages, they would not succeed.

6.2.2 The traditional design process

Open innovation and pivoting methods are the two main processes used when developing technologies and business models for rural markets through shared value. Shared value is an approach to innovation in which companies look for ways to grow and sustain their own businesses and create societal value by addressing society's needs and challenges. Open innovation is based around the free flow of knowledge

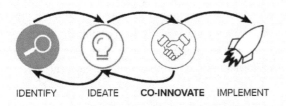

IDENTIFY   IDEATE   **CO-INNOVATE**   IMPLEMENT

Figure 6.1: The typical/previous smart village design process.

where both the giver and receiver have value to exchange [2]. Figure 6.1 provides an overview of the process.

The process begins with examining the villagers' pain points and connecting them with the right pain-relieving agents (industry partners). This step is called identify, and it is crucial to achieving the goal of making life better for the people living in rural communities.

However, these pain points must relate to a valid business opportunity to the providers such as startups and corporations. The second step in the process is to ideate, which is to help develop business models that optimize resource distribution and meet corporate objectives. The third step is to co-innovate, which includes business enterprises developing and delivering affordable technologies that rural villagers are willing to pay for. The iterative process is embedded in the co-innovate step. The iterative process uses training and awareness starting from the corporations, who then educate their fellows, who in turn train the smart village directors. The directors help their interns who then educate the villagers and collect data and feedback, which goes back to the refine technology offered by the industry partners. After identifying, ideating, and co-innovating, advances go through "proof of concept," which involves evaluating the sufficiency of the results to reiterate if needed; evaluation focuses on a corporation's business needs. Last, these ideas are implemented and scaled.

6.2.3 The WPI Ghana Project Center

The WPI Ghana Project Center alongside their local partners aim to push development design beyond its current framing by drawing on new design thinking, cross-cultural co-creation, and project-based learning, in ways that will reconceptualize the relationship between so-called "western" experts and the communities they hope to serve. In 2022, there are a total of seven IQP and MQP teams working on co-designing smart villages in Ghana upholding the project center's vision. The student teams are working with a variety of local partners varying from village chiefs, Academic City University College (ACUC) students, local professors, local business entrepreneurs, and more. To accomplish this, a new set of goals were redefined to prioritize community needs and to support what already exists in the villages. For these projects to truly benefit the community, it is necessary to disregard Western ideas of development and the typical process of creating smart villages and create a new process that upholds these reframed goals. Therefore, a co-design process was identified as a new methodology for creating an equitable new development process that puts generative justice at its forefront.

6.2.4 Generative justice

Generative justice is defined as the universal right to generate unalienated value and directly participate in its benefits that achieves a fair, sustainable exchange of value [4]. This ensures that communities of value generators have a self-sustaining path of circulation where value is not extracted and stays within the community. This concept allows equal benefits to everyone rather than one party having a status of more value for having more wealth. To break the pattern of alienated value an environment needs to be designed where the community circulates value. This ensures that value generators exchange value between peers instead of hierarchical structures. These goals are accomplished by working together with the local community; this is called co-design.

6.2.5 A co-design process

Co-design aims to build on the idea of peer-to-peer generation by ensuring that the design process functions as a free exchange of ideas. A detailed co-design map developed by WPI Professor Elizabeth Long Lingo was used as the inspiration for creating a new design process that upholds generative justice goals. The map is shown in Figure 6.2.

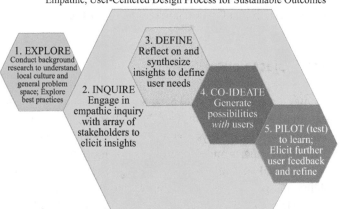

Figure 6.2: Steps of the co-design process (Lingo, ND).

As shown in Figure 6.2, the only part of the process outside the realm of the stake-holders – in our case the villagers – is the exploration step. This first phase in the co-design process includes conducting background research to understand local culture and the general to explore best practices. This step is crucial to gain the right amount of knowledge about the area before speaking with its people. Inquire is the second step, which is much more immersed with the stakeholders. This is done by engaging with stakeholders to elicit insights. The third step is to define the community needs, which ensures that the user needs to be established in the next step are accurate and defined by the community themselves. The third step is to define the community needs, which is done through reflecting and synthesizing the insights from the stakeholders. This def-inition step is still taking place within the realm of the community at stake, meaning consistent feedback and insights are being shared with the design team and the com-munity members/stakeholders. The fourth step in the process is to co-ideate, which means generating possible solutions to problems with the users. This step helps both parties take advantage of the strengths of each other and help one another think freely to create the most helpful solutions.

Lastly, to provide solutions, it is important to pilot and test designs to elicit fur-ther user feedback and refine what is needed.

6.2.6 Process improvement tools

To ensure that the design process used in the Ghana Project Center upholds the goals of generative justice, a set of Six Sigma and SEIPS process improvement tools were utilized in this project. A process improvement schematic was the main tool used to generate a new process, as described in Chapter 3.

6.2.6.1 Six Sigma

Six Sigma is a set of process improvement techniques used in more than 25% of For-tune 200 companies [5]. The Six Sigma methodology includes many process improve-ment tools utilized in this project such as SIPOC diagrams, process maps, and flow and swimlane diagrams.

SIPOC diagrams help define processes from start to finish and ensure the under-standing of existing processes. The diagram is used to gather information regarding the existing process conditions to assess and narrow the scope of the most important problems. In this project, SIPOC diagrams were used to understand the typical smart village design process to identify gaps and limitations. The tool is also utilized in the project team case studies to support future recommendations.

Flow diagrams represent a flow or set of dynamic relationships in a system. A com-monly used flow diagram in this project is swimlane diagram. This type of diagram develops understanding of who engages in what part of the process. Swimlane diagrams make it clear to their users what approach is being used in the process; top-down vs. bottom-up approaches become clearer through these diagrams. They also show the types of interactions going on between the different stakeholders within a process.

6.2.6.2 SEIPS model

To understand the context that occurs around the design process, a set of SEIPS tools were needed to consider the system as a whole. The SEIPS model is rooted in human-centered systems such as the creation of smart villages. It considers three major components of any system, plus the work system, processes, and outcomes as shown in Figure 6.3. This model dives into the major characteristics of each component and how they affect and interact with one another. Within every work system, SEIPS accounts for how the people involved interact with the tools needed, tasks assigned, and environments [6]. These components will then interact with the work process and finally these result in the work outcomes.

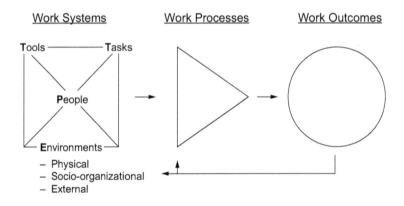

Figure 6.3: SEIPS model components and their interactions [6].

There are several SEIPS tools used in this paper to evaluate the case studies such as PETT scans, people maps, and outcomes matrices. A PETT scan summarizes the different components within a work system that include the people, environments, tools, tasks, and their interactions with one another. The PETT scan contains the barriers and facilitators of each of these components and their interactions. It is a flexible tool that can be used for intervention design to learn which factors to address when creating new designs, along with data collection and analysis. Lastly, PETT scans are great tools for understanding the priorities of the components within the work system against numerous factors. The second SEIPS model utilized was the people map, which represents the various people involved in a work system and how they interact with and relate to one another. The last SEIP's model used was the outcomes matrix, which identifies the various desired outcomes and whether they represent the project outcomes and goals. Considering co-design and generative justice it is important to consider conclusions for various stakeholders and the outcomes matrix is a great tool for that documentation.

Another important concept employed in this project is the concept of depth and breadth levers that are critical in transforming short-term, narrow-focus process improvements into long-term solutions [7]. The breadth and depth levers go hand in hand with the SEIPS models in which both methodologies state that the outcomes of a process do not solely depend on the process itself. These depth levers include "roles and responsibilities; measurements and incentives; organizational structure; information and technology; shared values and skills." All these aspects of a work system are crucial to take into consideration when reframing smart villages to fit our new set of generative justice goals.

6.3 Methodology

To achieve the project objective to design an improved development process addressing co-design and generative justice principles, I followed the continuous improvement approach outlined in Figure 6.4, which is an ongoing effort to improve processes through incremental improvements within an existing process [8]. This schematic was chosen due to its ability to be used in a variety of projects. The process includes enough detail to follow but is flexible enough to be altered based on the needs of each individual problem.

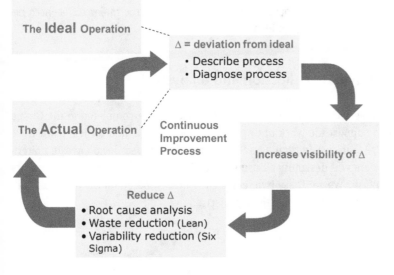

Figure 6.4: Process improvement schematic.

6.3.1 The traditional design process

The first step in my work was to explore the traditional design process, which in Figure 6.4 corresponds to "the Actual Operation." This step was accomplished by conducting background research and exploring contents discussed in "How to Create Smart Villages" [2], which laid out the current steps of designing smart villages. This helped me to develop a better understanding of the goals and motives behind creating smart villages and allowed me to pinpoint the differences between those goals and the reframed goals identified for the Ghana Project Center. I then developed a SIPOC diagram and swimlane diagram to summarize the information gathered and pinpoint the exact places that need improvement and those that do not.

6.3.2 Initial co-design process

Given that project teams were only familiar with co-design principles and were not introduced to a formal step-by-step design process, my goal was to create a formal ideal process that could guide project center teams in their work. This initial framing of the process incorporated co-design principles, representing the ideal process in Figure 6.4. I did this by thoroughly understanding Figure 6.2, which is the co-design process mapped out by WPI Professor Elizabeth Long Lingo, to determine an initial ideal design process based on generative justice goals. I also communicated with Professor Robert Krueger, the Director of the Ghana Project Center, who provided helpful co-design and generative justice resources. Laying out the initial design process was crucial to gaining feedback and insight from the project teams and further understanding the necessary changes. This initial co-design process helped me understand the major differences between the traditional and co-design ideas.

6.3.3 Case study data collection and analysis

Two case studies were chosen from the current projects being completed in the Ghana Project Center to support the work of this project, and an MQP and an IQP team were selected. The MQP team was designing a Stirling engine that uses local e-waste materials and the IQP team was designing a business model to create value for the community using available plastic waste. These two teams were chosen based on the availability of their members and the variability in their projects.

6.3.3.1 Data collection

Data was collected from these two case studies through interviews, observations, and informal group conversations. The initial meetings with the teams were held to understand their ways of going about their projects and how they were incorporating generative justice through their work. These meetings were essential for me to understand their goals, how they planned to achieve these goals, what generative justice looks like for each project, and where generative justice is lacking. By gathering this information, the goal was to compare with the initial ideal process to understand how to better fit and accommodate the needs of the different teams to accomplish generative justice goals.

Another source of data was observing team meetings with stakeholders and Ghanian partners. This helped me understand the types of interactions happening between all the different parties involved. In these observations, I sought to answer the following questions: how often were teams meeting with stakeholders? Who exactly did they meet with? What are the goals behind these meetings? Who initiated the ideas? And what are the limitations of these interactions? Lastly, I met with individual team members to gather feedback on the suggested new process. These discussions were helpful in making changes and revising the initial new process.

6.3.3.2 Data analysis

SIPOC diagrams, swimlane diagrams, and SEIPS models were used to analyze the data collected for each team. SIPOC diagrams helped me better understand each of the teams' processes and compare their work with the typical process. I then used all the data collected from the teams and created swimlane diagrams to help me better understand who did what in their processes and the sequence of the steps taken. I made use of SEIPS models, such as PETT scans, outcomes matrices, and people maps, to explore breadth and depth levers of the system as a whole and what might be needed to change the outcomes of the process. The PETT scan considers the full breadth of the work system [6]. Using this SEIP tool helped understand facilitators for the PETT scan's components and their interactions with one another. People maps were used as an addition to represent the people involved in the system and how they interact with one another. The last SEIPS tool employed was the outcomes matrix – one of the most important tools for data analysis. An outcomes matrix identifies the results of interest and whether they match the project's goals, or in our case the generative justice goals. These tools were all crucial to understanding and are important to think about when making changes to the initial process to achieve reframed goals.

6.3.4 Revised co-design process

Referencing the process improvement schematic in Figure 6.4, the data collection and analysis of the project teams were the main sources used to identify gaps, increasing the visibility between what the process is and how it should be. The last step was to try to reduce these gaps. After analyzing the data, I modified the initial process and mapped out a revised process.

Changes were made along the way, the data was collected and analyzed until the "actual operation" deviated less from the "idea operation" according to the process improvement schematic.

6.4 Results/discussion

In this chapter, the results of the project are presented. The chapter begins with an analysis of the traditional process, which highlights the key deficiencies in the traditional process from a generative justice perspective. Section 4.2 presents the results of the two student project case studies and the key observations from their analysis. Lastly, Section 4.3 presents the revised co-design process and the different components consisting of the work system, the process, and the outcomes that prioritize the goals of generative justice.

6.4.1 Analysis of traditional process and initial co-design process

The traditional design process discussed in Chapter 2 is described as the process of identifying the pain points of the villagers and the pain-point relievers, ideating the possible business solutions, co-innovating with the different stakeholders, collecting feedback from the villagers, testing proof-of-concept solutions, and finally, implementation. Figure 6.5, which shows the SIPOC diagram for the traditional process, highlights some of the key elements, including who is involved and some of the major outcomes from such a process. Industry partners, corporations, and smart village organizations are a few of the major beneficiaries of this process. These disproportionate benefits are shown in the outputs column; generating profit for industry partners by creating business opportunities, expanding markets and sales for global corporations, and finally it is stated that these smart villages will help villagers.

When used for designing many current systems, this process often removes value from a community by making them consumers rather than collaborators. The entities that do generate value – in many cases the business corporations working on these solutions – often create a system of injustice as observed in Figure 6.5. Although at first glance the traditional design process seems like an effective process for creating a smart village, this method puts villagers' needs last and goes through many major steps with-out including those who are most impacted by this work. The feedback collected from the villages is meant to help corporations understand what pain points will generate the most profit for these corporations [2]. The organizations participating in the smart vil-lage initiative will only target problems that provide profitable outcomes.

A good example of this is in the response to the opportunity to help millet farmers in India who were exploited by go-betweens and did not have the option to add value to their raw materials. Nestlé came to the rescue only after they confirmed that these challenges translated to a huge business opportunity [2]. The corporation stated that healthy nutrition is a core business to them and that organic, healthy millet bars would be a huge market within their scope.

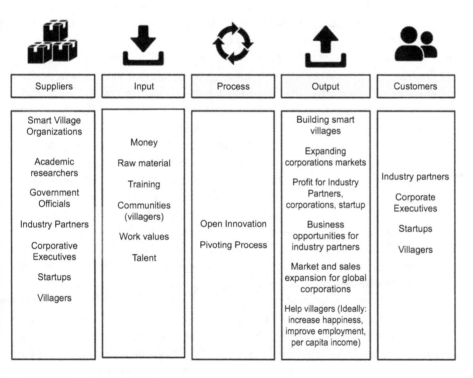

Suppliers	Input	Process	Output	Customers
Smart Village Organizations	Money		Building smart villages	
Academic researchers	Raw material		Expanding corporations markets	Industry partners
Government Officials	Training		Profit for Industry Partners, corporations, startup	Corporate Executives
Industry Partners	Communities (villagers)	Open Innovation	Business opportunities for industry partners	Startups
Corporative Executives	Work values	Pivoting Process		Villagers
Startups	Talent		Market and sales expansion for global corporations	
Villagers			Help villagers (Ideally: increase happiness, improve employment, per capita income)	

Figure 6.5: SIPOC diagram for traditional process.

Traditionally, smart village projects only address community needs that will benefit these large corporation's goals. Western nations and companies act in their best interests with-out adapting to different cultures and customs. This reinforces the colonial aspect of any project in which developed nations pursue the goal to "help" a developing nation. In many cases, instead of serving marginalized people, Western researchers and academics make their own assumptions about these communities based on a few – if any – encoun-ters with them (Smith, 2013). Development has come a long way, but without reallocating power back to countries healing from the impacts of colonization and allowing the most marginalized to make their own decisions, positive change will not happen.

The key issues with the traditional design process are not the specified process steps but rather the motivation behind them and the system in which the process takes place. Therefore, changing the goals behind the smart village development process is a first step to changing the results themselves. With the new goals of generative justice comes a new process, one that puts the villagers first and works hand in hand with partners on the ground to co-design innovative solutions to issues identified by the people. Generative justice ensures that value is generated within the community and is sustained through community members. The co-design process mentioned in Chapter 2 aligns with the goal of promoting generative justice through the work of the Ghana Project Center. Including the community in most of the design process ensures that the solutions being implemented are from and for the people themselves. It secures sustainable implementation and promotes self-sufficiency because the villagers are aware of the resources available in the community and are experts in the design process of maintaining these solutions overall.

Considering the co-design process map and generative justice definition and goals, an initial co-design process was developed for designing smart villages in Ghana and is shown in Figure 6.6. The smart village design team, which includes the WPI Ghana Project Center and students, should first talk to their Ghanaian partners. Communicating with the partners upfront is in place to discuss pain points, challenges, problems, community needs, and the resources available to outline what support is needed from the smart village design team. The next step is to conduct background research in such areas, which includes understanding local culture and exploring best practices. This step is equivalent to the explore step in the co-design process described in Chapter 2. Based on the empirical findings, the design team can brainstorm ideas to share with their stakeholders. These ideas will be discussed with the community leaders/stakeholders/partners for the people on the ground to understand the current ideas and provide sufficient input and feedback. After addressing the feedback, the team can work on designing solutions. Furthermore, to acquire funding for these projects, the design team will then pitch these ideas to grassroot organizations, NGO's, corporations, and the local government. After funding is settled, both the design team and partners will work together on creating solutions. This helps the community partners make sure that solutions are feasible on the ground and become experts on the design themselves. After the designs are built, they are ready to be piloted and tested. If this step is not successful in the eyes of the community, there is a need to reiterate and go back to brainstorming different solutions. This is called proof of concept.

Designs are then implemented and sustained through community contribution to continue the work.

Although the goal of this new process is to move away from the old process in terms of its final goals, we find that some steps used in this updated process are similar to the traditional one. This takes us back to the depth and breadth levers explained in Chapter 2. The depth levers include roles and responsibilities; measurements and incentives; organizational structure; information technology; shared values; and skills.

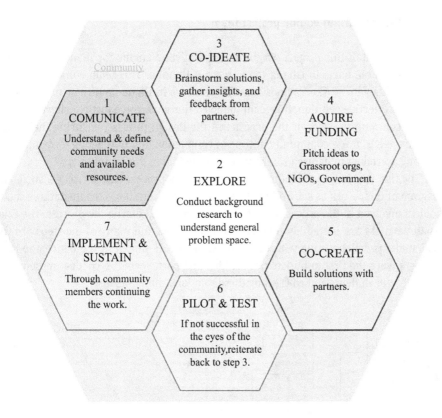

Figure 6.6: Initial new design process map.

This indicates that the steps in the process are not the only factors in a system affecting the outcomes of a process; rather it is the system as a whole.

6.4.2 Case studies and analysis

In the following section, I explore the two case studies chosen to be observed in this project. These case studies helped me determine how the actual process was working. In describing and evaluating the two project teams, identified the gaps needing to be reduced, which led to the improved co-design process. I utilized the SEIPS framework and focused on the two teams work system, their design process, and outcomes. I met with each team multiple times throughout the term to understand their work processes and goals with respect to generative justice principals. Through meetings and observing the teams' interactions with local partners, I created swimlane diagrams and utilized SEIPS tools such as PETT scans, people maps, and outcomes matrix to further evaluate their work in terms of the reframed goals.

6.4.2.1 Case 1: Stirling engine project team

The major goal of this project was to design a Stirling engine for manufacture using locally available parts in Ghana to charge a cellphone. This team consisted of four WPI students who worked alongside ACUC students in Ghana and their professor. To ensure an effective solution, this team followed this generative justice criteria: involving the community users of the product in the design process, ensuring the design suits the needs of the community at time of completion and in the long term, and last, creating a product to be used safely by any individual in the community.

The first step in understanding this team's design process was to create a swimlane diagram of their work as shown in Figure 6.7. This diagram shows the main three stakeholders involved in the process and the roles each played during the process. The Ghanaian partners are incredibly involved in this process and many of the steps are the WPI students collaborating with the ACUC students. As shown in the figure, the local partners were the first to initiate project idea and participated in co-ideating solutions with the WPI students and co-creating these solutions simultaneously.

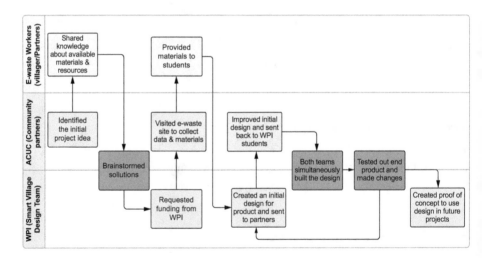

Figure 6.7: Swimlane diagram for the Stirling engine project.

The second step was utilizing the PETT scan and the people map from the SEIPS models, which provide an overview of the work system. The PETT scan includes the components of the work system that affect the result. Figure 6.8 includes the people, environments, tools, tasks, and their interactions. Creating the PETT scan helped inform me about the barriers and facilitators for each factor that can impact the outcome of the project. I identified these barriers and facilitators by observing team meetings and discussions with team members as well as the Director of the Ghana Project Center. As shown in Figure 6.8, most of the barriers are due to the project teams not being on the

Factor	Barriers	Facilitators
People		
– Students (ACUC and WPI) – Professors (ACUC and WPI) – e-Waste workers (Ghana and Worcester) – Ghanaian community	– Lack of time commitment from ACUC students due to different expectations	– Meetings with partners are productive – Positive teamwork between WPI students and ACUC students
Environments		
– Physical (Lab spaces in both WPI and ACUC) – Socio-organizational (WPI Project Center) – External (e-waste sites)	– Different time zone – Not being on the ground (distance)	– e-Waste sites (Ghana and Worcester) have been a big provider of materials – Background support provided from the Ghana Project Center
Tools		
– Knowledge – Communication platforms – e-Waste materials – Ideal Stirling engine	– Limited methods for contact – The ideal Stirling engine was difficult to get	– Access to an ideal Stirling engine provided a head start to the work
Tasks		
– Conduct research – Brainstorm ideas – Data collection – Design solutions – Test design – Implement solutions	– Limited data collection resources – Short time frame	– Brainstorming solutions with partners played a huge role in the project's success – After testing, feedback given by local partners helped reshape solutions and made more impactful
Interactions (among people, environments, tools, and tasks)		
– ACUC Professor provided the initial project idea – ACUC and WPI students visited e-waste site – ACUC professor and WPI students met e-waste workers – ACUC students and WPI students met and shared ideas – ACUC students collected data and shared with WPI students – e-Waste sites provided materials for project	– Miscommunication between WPI and ACUC students (people) due to limited ways of contact (tools) – Scheduling meetings (tasks) with different parties (people) is difficult – WPI students (people) do not have access to e-waste materials (tools) from the actual e-waste site (environment)	– Knowledge, information, and ideas (tools) are shared between different parties (people) – Combined meetings with students, e-waste workers, professors (people) are helpful and make sure everyone understands where the project is going (tasks)

Figure 6.8: Stirling engine project PETT scan.

ground in Ghana because of the COVID-19 pandemic. Not being physically onsite caused communication difficulties due to time zones, limited methods of contact, data collection shortages, inability to evaluate local materials for designs, and more. As for the facilitators for the Stirling engine team, they found interactions with local partners crucial to the success of their work, meetings, and teamwork between WPI students and ACUC students were productive, e-waste sites in both Ghana and Worcester were a huge help, and after testing on the ground, local partners provided feedback that reshaped the design and created more impactful solutions.

The exchanges between students and local partners are mapped out in Figure 6.9. Creating this people map helped show who was involved in the Stirling engine team's process and the types of interactions they had with one another. This information was collected through the meetings I had with the team and inquiring about how they worked with the local partners. As shown in Figure 6.8, the WPI students interacted with every local partner and other stakeholders such as the Ghana Project Center professors. Given the goal to co-design with local partners, it is best to have as many communication arrows as possible flow between everyone involved in the process. Figure 6.9 is a good example of such a well-rounded flow.

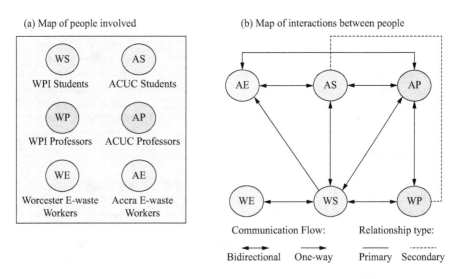

Figure 6.9: Stirling engine project people map.

Lastly, the outcomes matrix in Figure 6.10 shows the various effects of interest for this specific project in terms of generative justice. The outcomes matrix includes both proximal and distal outcomes. Proximal being the short-term goals and distal being the long-term goals.

		Outcomes for:		
		Ghanaian community	**Ghanaian e-waste workers**	**WPI students and professors (smart village organization)**
Proximal	**Desirable**	– Product can be used safely by any individual – Community can charge cellphones using the engine	– Ability to maintain the engine short term – Learn different ways to utilize e-waste	– Gain knowledge and experience in using reusable material for design and interdisciplinary work
	Undesirable	– Community members do not take advantage or are unaware of this resource	– Lack of familiarity with how the engine works and how to replicate it	– Not working alongside Ghanaian partners
Distal	**Desirable**	– Suits the long-term needs of the community – Creates business opportunities	– Ability to replicate the engine using e-waste materials	– Continued development of the engine for future projects
	Undesirable	– Community members do not find this solution essential to other priorities	– Difficult and expensive to replicate the design	– The design team has career goals inconsistent with community goals – Not having the community's best interest at heart

Figure 6.10: Stirling engine project outcomes matrix.

Given the COVID-19 pandemic and the inability to work on the ground, the WPI Ghana Project Center and students put in much effort into a co-design process that achieves generative justice goals. The Stirling engine team found similar difficulties to those experienced by the plastic recycling team, many to do with not being on the ground. Meeting with partners on the ground was crucial but hard to make it happen due to the time differences; these meetings helped ensure that everyone involved was on the same page and understood where the project was heading. This team said that they were constantly exploring and communicating with the partners on the ground that informed me the exploration step is not singular, but it is constantly occurring. Co-ideating with the partners on the ground was the most crucial step the team accomplished, especially because they were working hand in hand with the ACUC students, and the project could not have been done without the e-waste workers in Ghana.

6.4.2.2 Case 2: plastic recycling project team

This project aimed to tackle the problem of rural plastic waste management in the Eastern Region in Ghana. In partnership with numerous local chiefs, this team developed an actionable plan to establish a regional recycling partnership to coordinate collection, transportation, and sale of plastic waste. To do this, they developed a co-design framework that governed the design process of their proposed system to ensure that it is generativity just, culturally centered, and scalable. With these design principles established, they worked with entrepreneurs, local partners, and other stakeholders to determine management, funding opportunities, and supply chain logistics. They provided a data-driven feasibility report outlining the necessary stakeholder contributions to ensure the sustainability of their project. This was followed by an extensive discussion on the lessons learned over the course of the design process and to better prepare future partnerships for the challenges inherent in successfully co-designing cooperative businesses.

As shown in the swimlane diagram in Figure 6.11, the plastic recycling team consisted of three stakeholders: WPI students, local partners including village chiefs, and Ghanaian business entrepreneurs. The project idea was identified by the local partners, who brainstormed solutions with WPI students and provided feedback and insight before the WPI students went over any major steps.

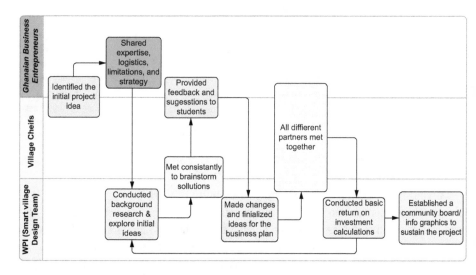

Figure 6.11: Plastic recycling project swimlane diagram.

Similar to the Stirling engine team, the plastic recycling teams barriers were mostly due to not being on the ground and the facilitators were due to working hand in hand with local partners, which is shown in the PETT scan in Figure 6.12. The people map shown in Figure 6.13 shows that this team worked with a variety of partners including villages chiefs, the former Ghanian ambassador to the United States, and local business entrepreneurs. The team informed me that the meeting they had with multiple stakeholders all together was essential to the success of their project. Figure 6.14 shows the outcomes matrix for the plastic recycling team, which shows their desired goals considering generative justice principles.

Factor	Barriers	Facilitators
People		
– Students (ACUC and WPI)		– Brainstorming
– Village chiefs (Denase, Batabi,		environment worked well
Tumfa, Abompe)		– Students and partners
– Ghanaian business representatives		together reached
– Former Ghanaian ambassador		reasonable solutions
– Ghanaian community		faster than assumed
– Student who previously worked		– Denase chief helped
with the chief on a plastic project in		students contact more
Ghana		local partners
Environments		
– Physical (WPI spaces)	– Different time zone	
– Socio-organizational (WPI Project	– Not being on the ground	
Center)	(distance)	
– External		
Tools		
– Knowledge	– Limited ways of contact	– Readily available plastic
– Communication platforms		in Ghana
– Plastic waste		– Knowledge shared by
		village chiefs and
		business entrepreneurs
		was essential to the
		teams' work
Tasks		
– Conduct research	– Brief time frame	– After testing, solutions
– Contact partners		were completely
– Brainstorm ideas		reshaped, and the
– Data collection		process went more
– Design solutions		smoothly the second
– Return on investment calculations		time
– Implement/sustain solutions		

Figure 6.12: Plastic recycling project PETT scan.

Factor	Barriers	Facilitators
Interactions (among people, environments, tools, and tasks)		
– WPI students met consistently with the Chief of Denase. – WPI students met with chiefs of Batabi, Tumfa, and Abompe – WPI students met with plastic business entrepreneurs – WPI students met with the former ambassador of Ghana to gain credibility – WPI students, village chiefs, and the former ambassador met all together – WPI students contacted ACUC students to make educational resources and running surveys – WPI students met with students who worked on plastic recycling projects before to gain insight	– Scheduling meetings (tasks) with different parties (people) all together on one platform (tools) is difficult	– Knowledge, information, and ideas (tools) are shared between different parties (people) – Combined meetings with students, chiefs, and Ghanaian ambassador, (people) are helpful and make sure everyone understands where the project is going (tasks)

Figure 6.12 (continued)

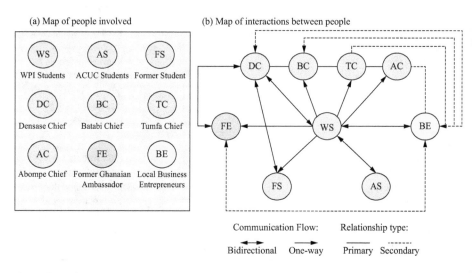

Figure 6.13: Plastic recycling project people map.

		Outcomes for:		
		Ghanaian community	**Village chiefs and plastic entrepreneurs**	**WPI students and professors (smart village organization)**
Proximal	**Desirable**	– Utilizing available plastic waste for the benefit of the community	– Ability to maintain the business plan short term – Learn different ways to utilize plastic waste	– Gain knowledge and experience in using reusable material for design and interdisciplinary work
	Undesirable	– Plastic recycling project only benefits certain people within the community	– Inability to maintain the business plan	– Not working alongside Ghanaian partners
Distal	**Desirable**	– Suits the long-term needs of the community – Creates business opportunities	– Establishing a joint board with different representatives who will sustain the project – Utilize infographics made by students and educate the community about plastic waste	– Continuation and development of the business plan for future projects
	Undesirable	– Community members do not find this solution essential in comparison to other priorities	– Conflict within the different business partners	– The design team has career goals inconsistent with community goals – Not having the community's best interest at heart

Figure 6.14: Plastic recycling project outcomes matrix.

During my discussion with the plastic recycling team about their process, they mentioned that step two, Explore, in the initial design process shown in Figure 6.6 was the most useful. This was because the students are outsiders when it comes to the community especially working remotely, so they found communicating with the partners on the ground essential. They noticed that when they went through the complete process many times, they would work with their partners, gain insight and feedback, then go back and change the work. This team found the distance, due to working virtually, a setback to the success of their project during the co-ideate and co-create steps. Solutions took more time to complete because of the distance. They also brought up that technology was a big setback, that making group calls work was difficult. They observed that the people on the ground used different platforms of communication and it was hard to reach all the stakeholders at once. However, when meetings did happen, they were very helpful, and brainstorming went well. Meetings were limited, and the students could not communicate with all the different stakeholders as they wished to. This team did not need to acquire funding as a complete step; they stated that their project's goal was to create something that the community can then use to generate value, which meant funding was not a major step in their work. Rather, the goal was to ideate and co-create solutions together.

6.4.3 Revised co-design process

Utilizing the SEIPS methodology and depth and breadth levers, I identified a complete system that should be put in place to ensure generative justice goals are met. This is because the process itself is not the only thing affecting what comes out of it and to achieve the generative justice goals, there is a need to consider the system as a whole. This system consists of three main elements: the reframed goals and desired outcomes, the process, and the work system.

6.4.3.1 Reframed goals and outcomes

The goals and desired outcomes of the new process have been identified in previous chapters in this book. The goal of achieving generative justice through co-design means the smart village organization needs to play an empowering role and work with the community on creating projects. There should not be any power balance or assigned roles such as the "helpers" and "receivers." Everyone in this system is working together toward one goal. Ideally, the community members are completely empowered to sustain these projects after the smart village organization moves onto another project.

6.4.3.2 The new improved process

Figure 6.15 shows the improved finalized design process map. Although these steps do not differ drastically from the traditional process, the system around the process is what helped accomplish the desired goals. This process would be a success for many projects but if the steps are not surrounded by the right people, environments, motivations, and more it would not accomplish generative justice goals.

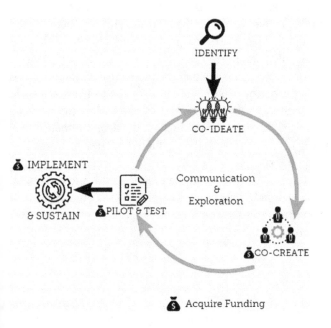

Figure 6.15: Improved co-design process.

I used the two case studies to gain insight on what worked well in the initial idealized co-design process and what still needed improvement to map out the final improved process shown in Figure 6.15. The first step both teams took was clear: identifying the problems was essential to starting any project. This step also was first in the traditional design process; however, the motives behind it are different now. The reason identifying the pain points is needed within a co-design process is to listen to the community and allow them to identify what they need and want. During this time, and going forward, the team should be exploring while communicating with partners best solutions, the culture, the available resources, and more. This step is always happening and is necessary for a true co-design process. Funding is also not always needed for every project to be successful. WPI teams relied on the university resources so they did not need to acquire funding; however, if they did, it would have been incorporated in any of the co-create, testing, or implementation steps.

Co-ideating is the next actual step after identifying the problems. Co-ideating looked different in these projects due to the lack of contact with the partners given the COVID-19 pandemic. However, the teams worked around this by brainstorming with their partners when they had the chance to and if it was not possible to schedule a meeting, they would gather feedback and insight on these ideas as soon as they could and would not go forward with any plans without their partners' approval. Co-create is the third step. This was simultaneously done by teams at WPI and the teams on the ground creating the solutions. For example, the Stirling engine team were building the suggested engine as the ACUC students were building it onsite. This way not only are the partners able to understand and learn how the product works to provide insight and feedback but they are also able to replicate it in the future when the WPI students complete their projects. The fourth step is to pilot and test these created solutions. This is ideally tested on location by the partners and if not viewed as helped or a success in the community's eyes, then the process is reiterated, and we are back to the co-ideate step. This reiteration process was proved successful by both the teams. Although they did not reach the testing step in their projects, they gathered enough feedback and could do basic calculations on the success of the solutions to reiterate when needed. Lastly, all these solutions will be implemented, and the goal is for them to be sustainable. Generative justice and co-design if done correctly work on ensuring projects are sustained through the community especially because they are the experts on these projects and can make them grow.

6.4.3.3 The work system

To achieve reframed goals through the improved process, it is important to create the right work system surrounding it. This includes the people, their interactions, the environments, resources, and the tasks taken upon the different stakeholders involved. It is essential that everyone involved in the process is aware of the reframed goals and how that will affect their work. The motivation behind what problems are chosen to work on and who chooses these problems is a substantial change compared to the traditional design process. This work is highly dependent on the Ghana Project Center because they oversee assigning projects to students and are in most contact with local partners on the ground.

Specific recommendations for each element of the work system include:

- Tasks: (1) It is important to ensure that teams are consistently communicating with their partners. This is easier said than done especially when working remotely in two different countries. However, this can be accomplished through the Ghana Project Center by (a) planning early methods of communication with WPI students and local partners, gaining prompt access to contact information of partners ensures that co-design is being implemented from the get-go of the projects and (b) ensuring that the partners are the ones coming to the WPI Project Center and not the other way around. (2) Another suggestion is making sure that the local partners are supported after the completion of the projects to sustain and continue the work on their own.

– Tools: (1) Ensuring that WPI students have access to similar tools as those on the ground to get the most accurate results. For example, the Stirling engine team struggled with using e-waste sites from Worcester because they do not completely represent the material found in the e-waste site in Ghana. (2) Being informed that local partners use different communication apps than those used by WPI students was a helpful start for WPI students.

– People: (1) The Ghana Project Center needs to ensure that there are enough local partners who can work with students on projects every step of the way. (2) Continue the preparation before the start of the projects such as learning cultural differences, local language, and more. This is helpful even in a remote setting because it allows students to know how to approach meetings with partners virtually. (2) Ensuring that the goals of generative justice and co-design are aligned with all the different stakeholders.

– Environment: (1) Ensuring that there are people on location who oversee data collection especially when students are in Ghana. For example, it was difficult for the Stirling engine team to evaluate their engine in Ghanian temperatures when they needed access to temperature data from partners on the ground. (2) Support mechanisms like the weekly ground progress reports and class discussions were helpful.

6.5 Conclusion and recommendations

Smart development is a movement that has fraudulently branded itself as a way to make the world a better place. It has been used as a tool for more privileged nations to gain economic growth and power. Although it is on a journey of progressing and considering the quality of life of the developing nations, there is still a long way to go when it comes to erasing its history and impact of being a postcolonial tool to colonize nations in the name of "helping" them. The WPI Ghana Project Center is wary of this issue and is working toward changing the intended goals for these types of projects and co-designing with the community to achieve generative justice goals. In 2022, seven WPI teams worked on developing smart village projects in Ghana; however, the design process to guide there was not formally defined. Traditional development processes for smart villages prioritized the needs of the corporations involved and viewed villagers needs as secondary.

This project focused on developing a new design process based on co-design principles to create smart villages that are self-sufficient and self-sustained, and generate value for those living in the community. A continuous improvement approach was used. After evaluating the traditional design process and suggesting an initial improved process, I used two projects in the Ghana Project Center as case studies to understand what was working well and what needed to change. Finally, an improved process was suggested that consists of five major steps: identify, co-ideate, co-create, pilot and test, and lastly, implement and sustain. These steps are crucial to the success of these projects but achieving the desired goals requires looking at the entire system. The work system included the people, environments, tools, tasks, and their interactions. Defining aspects of these work system components alongside the new co-design process can support the achievement of the desired generative justice goals.

Finally, I would recommend addressing several aspects of the work system to support the design process and the overall goal to achieve generative justice. Please see the following.

– Tasks: Constant communication between stakeholders is key. This can be done by early planning of communication methods and gaining access to contact information of local partners before the start of the project.
– Tools: Gaining access to similar tools as those on the ground is important in ensuring the designs created are successful in the local environment.
– People: Having the right people on location is the first step to success. It is also important that students are learning about cultures and language differences before starting their projects. Also, ensuring that students understand the non-Western design methodologies.
– Environment: In a virtual setting it is important to acknowledge the different environments the designs will be implemented in which is why having people in charge of data collection on the ground is crucial to the success of these projects.

Many of these recommendations would be much easier to implement if the Project Center were open in Ghana. In conclusion, the new system and process focused on co-design with concrete steps to achieve generative justice can be used by future project teams.

References

[1]   Escobar, A. (2014). Development, critiques of. In *Degrowth*. Routledge, pp. 57–60.
[2]   Darwin, S., Fischer, W., & Chesbrough, H. (2020). *How to Create Smart Villages: Open Innovation Solutions for Emerging Markets*. Peaceful Evolution Publishing.
[3]   Darwin, S. (2018). *The Road to Mori*. Van Haren Publishing.
[4]   Eglash, R. (2016). An introduction to generative justice. *Teknokultura, 13*(2), 369–404.

[5] Jones, E. C., Parast, M. M., & Adams, S. G. (2010). A framework for effective Six Sigma implementation. *Total quality management*, *21*(4), 415–424.

[6] Holden, R. J. & Carayon, P. (2021). SEIPS 101 and seven simple SEIPS tools. *BMJ quality & safety*, *30*(11), 901–910.

[7] Hall, E. A., Rosenthal, J., & Wade, J. (1994). How to make reengineering really work. *McKinsey Quarterly*, 107–107.

[8] Soković, M., Jovanović, J., Krivokapić, Z., & Vujović, A. (2009). Basic quality tools in continuous improvement process. *Journal of Mechanical Engineering*, *55*(5), 1–9.

Index

https://doi.org/10.1515/9783110786231-008

Printed in the USA
CPSIA information can be obtained
at www.ICGtesting.com
LVHW080743180424
777763LV00005B/424

9 783110 786217